The Boy Who Invented Television

A Story of Inspiration, Persistence and Quiet Passion

Paul Schatzkin

The Boy Who Invented Television

TeamCom Books
Silver Spring, MD
www.TeamComBooks.com

Credits

Publisher:	Bruce Fries
Editor:	Chris Roerden
Proofreader:	Susan Higgins
Cover Design:	Kathi Dunn
Composition:	Bruce Fries
Author Photo:	Ken Gray

Notice of Rights

Publisher's Cataloging-in-Publication Data

Schatzkin, Paul
    The boy who invented television : a story of inspiration, persistence and quiet passion / Paul Schatzkin
        — 1st ed.
     p. cm.
     Includes bibliographical references and index.
     LCCN:
     ISBN: 1-928791-30-1

     1. Farnsworth, Philo Taylor, 1906 – 1971. 2. Television History. 3. Inventors—United States—Biography.
     I. Title.

     TK6635.F3.S39 2002   621.388'0092

Text set in Garamond Lite and Congress Sans.

Printed and bound in the United States of America
9 8 7 6 5 4 3 2 1

For Philo III, Donald, Michael,
. . . and Harvey

Acknowledgments

Many people have contributed to the research and writing that are synthesized in these pages. There is much more to say about the process that has produced this book, but for now I wish to express my profound gratitude to: the entire Farnsworth family, beginning with Elma G. "Pem" Farnsworth, and including Philo T. Farnsworth III, Kent and Linda Farnsworth, and Skee and Rose Farnsworth. Others who have contributed along the way: Bob Kiger, Georja Skinner, George Everson, Nick DeMartino, Richard Hull, Bruce Fries, and Chris Roerden. And of course my wife, Ann, who along with me has learned that although writing does not entail much heavy lifting, there is nevertheless a burden in words.

Website

For more information about Philo T. Farnsworth, visit "The Far-novision" on the World Wide Web: *http://farnovision.com*.

Contents

"Discovery consists of seeing what everybody else has seen, and thinking what nobody else has thought."

—Albert von Szent-Gyorgyi

The Second Case

Philo T. Farnsworth with the Fusor Mark I in 1960

"My father had two major cases in his life. The first
one should have brought him the ability to deliver
the second, and it did not."
—Philo T. Farnsworth III

"You just don't get it," Phil said, shaking his head. "I've
given you all you need to finish the patents. Now I'm going home and get drunk."

With that, Philo T. Farnsworth calmly collected his papers,
closed his briefcase, rose from his chair, and left the meeting,
turning his back on the single most important invention of the
20th century—if not the whole second millennium.

Left behind to "finish the patents" were Farnsworth's patent
attorney, his mathematician, and his principal corporate benefactor. The idea that his colleagues just didn't "get" was Farnsworth's
path to the Holy Grail of modern science, his answer to the riddle
of controlled nuclear fusion.

Fusion, if it can ever be perfected, offers the promise of a clean, safe, and virtually inexhaustible source of energy. In a world beginning to choke on fossil fuel emissions, Philo Farnsworth had devised a pollution-free process for extracting the energy that binds atomic nuclei.

The fuel for fusion is an isotope of hydrogen, which is easily and cheaply extracted from water. While the industrial world was aggressively drilling for enough oil to quench the planet's insatiable thirst for energy, Philo Farnsworth had discovered a source of power as vast as the oceans themselves, where each gallon of seawater would produce as much energy as hundreds of gallons of gasoline.

In Farnsworth's fusion-powered future, a dynamo the size of a volleyball would produce as much electricity as all the generators at Niagara Falls—and less exhaust than a lawnmower. In a world that would soon experience gut-wrenching disruptions of its oil and gas supplies, Philo Farnsworth had found a way to light a city the size of New York for less than a dollar a month.

While the rest of the scientific community was sinking into the quagmire of its deadly fascination with nuclear fission—a process that produces terrifying levels of radiation and waste products that remain hazardous for hundreds of thousands of years—Philo Farnsworth had devised a process that would produce nothing more threatening than helium—a harmless, inert gas.

For Farnsworth personally, fusion was much more than a source of clean power. A fusion-powered engine would provide him with the means to fulfill his lifelong dream of traveling through outer space, in much the same way that internal combustion engines provide the means for traveling around the surface of the Earth. A full decade before the first man left his footprints on the Moon, Philo Farnsworth had found a way to travel among the stars.

In nature, we are surrounded by fusion reactors; the nearest one, the Sun, is a mere 93 million miles away. Every star in the heavens is a deep-space fusion reactor: a self-contained ball of gas and cosmic fire. For nearly fifty years, scientists have attempted to replicate the process here on Earth, trying to create what amounts to a miniature, synthetic star. But therein lies the riddle: how do you bottle a star? To a culture obsessed with television and movie stars, Philo Farnsworth offered an artificial sun.

Farnsworth had invented a device he called the "Fusor" that synthesized all the unique insights he had accumulated in more than forty years on the edge of the "invisible frontier" of the 20th century. To perfect his invention, he just needed to configure the Fusor in such a way that the reaction would continue on its own—just as an automobile engine keeps running after the starter motor is disengaged. During a number of experiments conducted in the early 1960s Farnsworth's co-workers claimed to have witnessed just such an occurrence.

When he walked out of the patent meeting that summer afternoon in 1965, Philo Farnsworth had made more progress toward controlling nuclear fusion than anybody before or since. He had come tantalizingly close to lighting "the fusion torch." Yet, despite his impressive progress in the laboratory, his co-workers still did not truly understand his theories, which often skated beyond the edge of generally accepted physics. It didn't help that no less an authority than Albert Einstein had endorsed Farnsworth's ideas.

Perhaps what Farnsworth's colleagues didn't "get" was the mounting cost of research for a branch of science that they regarded as well outside the scope of their corporate mission. Like it or not, Philo Farnsworth was an employee of the International Telephone and Telegraph Company. ITT was a multinational telecommunications conglomerate, and most decidedly was *not* in the nuclear energy business. The fusion research was only reluctantly funded.

Filing patents was no small part of ITT's strategy. Once the company owned patents on the Fusor, it intended to seek government funding for further research, rather than continuing to spend funds from its own corporate treasury. So in the summer of 1965 Farnsworth was asked to assemble his team and draw up the patents that would control the art and science of harnessing nuclear fusion—patents that ITT would own, and for which the government would hopefully fund further research.

Farnsworth was leery of the process. He didn't think the people around him understood enough of the underlying concept of his invention to compose enforceable patents.

As the work sessions began, Farnsworth found some small encouragement. It looked as if his colleagues were finally beginning to "get it." And then, at that critical moment late in the summer of 1965, as he was explaining a basic principle that made the Fusor work, it suddenly became evident to him that all his patience and persistence had been for naught. Something was said that crushed Farnsworth's passion and shattered his faith once and for all. At that instant he realized that his dream was not going to become a reality, at least not in his lifetime.

Things might have turned out very differently if Farnsworth had received the recognition and independence he deserved from his first major invention. You see, when Philo T. Farnsworth was still a teenager, he invented something called "television."

That television ever *was* an invention comes as a surprise to most people, a fact that seems tragically ironic, because invention plays such a big part in the American legacy. We all know the era of electrical communications began when Samuel Morse tapped out "What hath God wrought?" on the first telegraph. The recorded music business began when Thomas Edison recited "Mary Had a Little Lamb" and heard it played back moments later from a tinfoil drum. The telephone arrived when Alexander Graham Bell spilled some acid on his pants and shouted, "Mr. Watson, come here, I need you!"—and Mr. Watson heard him on a contraption in another room. The motion picture industry began when Edison filmed a sneeze.

Television represents the culmination of all the inventions that went before it: the marriage of movies and radio—sight and sound merged with the electromagnetic spectrum. Television was the crowning achievement of an age of invention. But who among us can name the man who invented it?

Video, in all its forms, is the most pervasive medium ever conceived. It's not just television, which is so omnipresent that you can't even wait in an airline terminal today without being compelled to watch CNN or a soap opera; it's computers, too, which have become a mass medium in their own right. Nearly every computer in the world uses video as its primary display device.

Yet how many of us whose lives are shaped by these devices have any knowledge of the unique individual who delivered them to us?

The corporate doctrine handed down over decades by the communications industry would have us believe that television was far too complex to have been "invented" by any single individual, working alone, in a garage perhaps, in the manner of Edison and Bell (or Hewlett and Packard, or Jobs and Wozniak). The industry would rather we believe that the medium *evolved* over a period of time, finally emerging whole in the late 1940s from the great laboratories of the industrialized world—just in time for Uncle Milty, Marshal Dillon, and Lucy. Otherwise, virtually no folklore is associated with the origins of television. This void in our popular mythology is unfortunate, because in fact, the story of the invention of television is one of the most fascinating stories of the 20th century... and features one of the era's most intriguing and enigmatic characters: Philo T. Farnsworth.

ONE

This Place Has Electricity!

The Farnsworth homestead near Beaver Creek,
Utah where Philo T. Farnsworth was born in 1906

"The reasonable man looks at the world around him
and tries to adapt himself to it. The unreasonable
man tries to adapt the world around him to himself.
All progress, therefore, is dependent on the
unreasonable man."
—George Bernard Shaw

The story of television begins in Rigby, Idaho, in the spring of 1918, as a small wagon train reached the crest of a hill overlooking the Bungalow Ranch, a humble, turn-of-the-century homestead named for two small cabins that dominated the landscape. The family of Lewis and Serena Farnsworth and their five children were about to arrive at their new home, after an arduous journey over the mountains from their native Utah.

Seated at the reins of one of the three covered wagons was the oldest child, Philo, age eleven, named after his grandfather,[1] who'd come west to settle the Salt Lake Valley with Brigham Young. While the adults in the wagons around him commented on the contours of the countryside and the promise of fertile soil, young Philo noticed one detail that the rest of the family missed: strung between the buildings below he could see *wires*, and shouted excitedly, "This place has electricity!"

With this observation, the family left the ridge and began their descent into a new life on the frontier of the 20th century. Philo T. Farnsworth was about to come face-to-face for the first time with the mysterious force he had only heard about; that invisible power that could drive great machines, carry voices over wires, and turn darkness into light. Though he was about to encounter electricity for the first time at age eleven, he would prove to be one its great masters before he was twenty-one.

The Farnsworth family was a poor but hard-working clan who had moved so many times in search of better land that the siblings often joked about how, once the horses were hitched to the wagons, the chickens and pigs would just lie down and put their feet in the air to be tied. This time, Lewis was bringing his family to Idaho to join forces with his brother Albert, who had enticed them with rosy stories about the rewarding future that awaited them—cultivating two-hundred-plus acres of sugar beets, potatoes, and hay in this rugged Idaho valley.

Traveling in three covered wagons, it had taken the family more than five weeks to cross the terrain from southern Utah. Lewis and Serena drove the lead wagon. Along with the younger children, the wagon was stuffed with mattresses, homemade quilts, dishes, and other fragile household effects. Philo's older half-brother, Ronald, and his wife drove the second wagon, carrying the stove and kitchenware. Philo drove the third wagon, carrying crates of chickens and piglets and farm tools. The little caravan was followed by several cows and extra horses.

Besides farming, Lewis Farnsworth often supplemented the family's income by taking on freighting jobs, using his horse-drawn wagon to haul produce and merchandise from the terminus of the railway to remote towns and settlements. Philo had a close relationship with his father, who often took him on his freighting trips. Though Lewis was not a schooled man, he had done all he could to educate himself, reading as many books as he could get his hands on—a trait that his oldest son was quick to emulate. On their freighting expeditions, Lewis and Philo often spent their nights under the open sky, and Lewis taught his son how to recognize the constellations and track the movement of the planets across the heavens. One time, Philo picked up a

Philo on the farm in 1922

long stick and aimed it at the sky, wishing that he could touch a star. Seeing the dreamer in his son, Lewis often cautioned him, "It's okay to keep your eyes on the stars, son, but keep your feet on the ground."

Before their move to the Bungalow Ranch, the family had little contact with the evolving world of machines and gadgets. The one modern luxury they did have was a hand-cranked Gramophone, and music was very much a part of their lives. Whatever else they knew of the changing world around them they learned from the Sears & Roebuck catalog, for so many families the great "wish book" of the day. The family was too poor to actually buy anything from the catalog, but the illustrations must have sparked young Philo's imagination about the world of science and invention. He had read about Erector sets, electric trains, and the small motors that ran them. At one point he had even thought about how he might make his own electricity, if that's what it would take to experience the stuff firsthand.

Once the family was settled in its new home, the budding boy electrician turned his attention to the ranch's Delco power system. He carefully watched the adults who operated the system until he learned all he could about how it worked. The power system was an indispensable part of the farm operations, providing power for the granary as well as lights for the house. There was just one minor problem—the system frequently broke down.

On one such occasion, repairmen came to the ranch to get the system going, and Philo tried to get himself as close to the action as possible, peering over the adults' shoulders, forcing his way between them as they hovered over the broken down generator. When he noticed that they were using heavy oil to lubricate the system, he tried to pull his father aside to tell him that it was the wrong kind of oil, but the repairmen managed to get the system running again, and the boy's warnings went unheeded. No sooner had they left the ranch than the generator conked out again.

This time, Philo would not take "no" for an answer. He told his father, "If you'll let me, I'll get it started again. I know what's wrong with it." As his elders all stood around in amazement, Philo very carefully disassembled the generator and meticulously cleaned each gunked-up part with kerosene. When he put it all back together, it started right up and ran smoothly, much better than the supposed experts had left it. After this episode, Philo was officially instated as the chief engineer of the Bungalow Ranch, and the electrical system became his own very special domain.

With encouragement from his father, it wasn't long before Philo found more uses for his invisible new friend. Among his chores on the farm was to turn the handles on his mother's washing machine, a monotonous task he found terribly boring. Scattered around the grounds, in a pile of junk that the previous tenants had left behind, Philo found a burned-out electric motor. He found some wire and rewound the armature, then connected his new creation to the washing machine—and was done with his handle-turning chores forever. The motor worked so well on the washing machine, he then adapted it to his mother's sewing machine—the first electric sewing machine she'd ever had.

The time he saved by automating his chores, Philo spent thinking about better things. The family had no money for books, but in an attic loft above one of the cabins, he discovered a treasure trove of magazines left behind by the previous occupants of the ranch, with titles like *Popular Science,* and Hugo Gernsback's *Science and Invention*. The loft quickly became his secret hide-away, his own private library. With each new page, the young boy's imagination became fired by stories of science. He read about Edison, Bell, Marconi, DeForest, and the other modern-day sorcerers who explored these hidden frontiers. To Philo, inventors of all kinds seemed to possess a special power that allowed them to see deep into the

Lewis Farnsworth

mysteries of Nature and use her secrets to ease the burden for all mankind. He confided to his father his own heart's desire: that he, too, had been born an inventor.

It was in his attic loft, among the discarded electrical magazines and science journals, that Philo also encountered the controversial ideas of an obscure German patent clerk named Albert Einstein, who had set the scientific world on its ear in 1905 by reconciling conflicting theories about the properties of light with his "Special Theory of Relativity." More recently, in 1915, Einstein had published a second groundbreaking paper, his "General Theory of Relativity," in which he introduced even more unorthodox ideas about gravity and the structure of time and space. In this later work, Einstein postulated that space was like an elastic fabric that could be contorted by gravity. This notion of a stretchable universe was considered heresy in the fixed, mechanical world of Newtonian physics.

Einstein had encountered tremendous resistance to his radical theories among the scientific establishment, but there was no such difficulty for an aspiring young scientist in the hills of Idaho whose imagination was not saddled with any preconceived notions of the workings of the physical universe.

Philo's interest in Einstein's universe deepened during a visit to a neighboring ranch. There, he happened upon a newspaper article about an international expedition that had sailed to the coast of Africa in May 1919 to test Einstein's theories by observing a solar eclipse. Led by the British astronomer Sir Arthur Eddington, the expedition arrived in Africa at a time when Europe was still reeling from the devastation of the Great War, yet scientists from both Allied and Axis nations had teamed up for the mission.

Eddington and others reasoned that if space could be stretched, as Einstein suggested, then light also could be stretched, or curved, by the presence of a large gravitational mass like the Sun.

When the eclipse had darkened the West African sky, Eddington's team photographed starfields around the Sun that were not visible during normal daylight. When these photos were compared with photos of the same starfields taken when the Sun was not present, Eddington found a difference in the apparent location of the stars—just as Einstein had predicted.[2]

The Eddington expedition was recorded as a defining moment in 20th century science. It was also a defining moment for a young boy reading about it in a borrowed newspaper on a remote farm in rural Idaho. Though he was not yet a teenager, what Philo was learning of Einstein's theories resonated instinctively with his inquisitive intellect. How exciting to read that the expedition had proven Einstein's assumptions about the fabric of the universe were correct.

Philo was even more inspired to read that the expedition to Africa included scientists from nations that were so recently at war with each other. He told his father that if scientists from warring nations could put aside their differences for the sake of discovery, then a scientist was a good thing to be.

With the fire of discovery burning in his young soul, Philo established a rigorous routine that would serve him the rest of his life, rising every morning at four, using the time when the house was quiet for his studies before starting his daily chores at five. After breakfast, he'd get his horses together to take the children to school—in a wheeled wagon in the warmer months, and in a sleigh during the winter, when the temperatures dropped as much as forty degrees below zero. Long hot summer days he spent tending the fields, planted mostly in hay and potatoes. Riding his horse-drawn mowing machine gave Philo plenty of time to think about the things he was reading. Nights he hurried through the last of his chores in order to get back to his attic hideaway, where he consumed anything about electrical science that he could get his hands on.

One noteworthy night, Philo turned the pages of one of his magazines and encountered an idea that resonated with a chilling premonition of his own future. In an article about "Pictures That Could Fly Through the Air," the author described an electronic magic carpet, a marriage of radio and movies, that would carry far-off worlds into the home in a simultaneous cascade of sight and sound. Philo was instantly captivated by the idea. He reread the article several times, convinced that he had stumbled onto a challenge that he was uniquely equipped to solve.

When Philo determined to learn everything he could about the subject, he stepped into a Jules-Vernian world where scientists were trying to convert light into electricity with the aid of whirling disks and mirrors. During the first two decades of the 20th century, experiments with "vision by radio" drew largely on the technology of the day, as inventors and engineers tried to literally blend the mechanical contrivances of motion pictures with the electrical properties of radio. The resulting contraptions. were given fanciful names such as "radioscope," "teleramophone," "radiovisor," "telephonoscope," and finally "television."

The latter term is attributed by some to a Frenchman, Constantin Perskyi,[3] by others to Hugo Gernsback, whose first use of the term appeared in the early 1920s in his *Science and Invention* magazine—no doubt one of the magazines young Philo was reading that memorable night in his attic lair.

Mankind's eternal desire to "see over the horizon" found its first practical stirrings with the discovery that tiny electrical currents are generated when light is shined on certain substances.

This discovery is most often attributed to two English telegraph engineers, Joseph May and Willoughby Smith, whose 1873 experiment with selenium and light gave future inventors a way of transforming images into electrical signals. Over the next two decades, this "photoelectric effect" was also observed by Heinrich Hertz, one of the early pioneers of radio. Other scientists, most notably Max Planck and J.J. Thompson, added further insights on the subject. But it was not until Albert Einstein arrived on the scene in 1905 that the phenomenon was fully articulated and quantified.[4] Einstein's paper on the photoelectric effect supplied the fundamental math that future researchers would need to realize mankind's "most fanciful dream."

Armed with their understanding of this photoelectric effect, the earliest television experimenters surmised that an image would have to be disassembled into its component parts of bright and dark elements. These individual picture elements could then be converted into an electrical current, the strength of which would fluctuate in accordance with the brightness of the picture elements. This current could be transmitted over wires or through the air, and the image would be reproduced on the receiving end by reassembling the original picture elements in precisely the same sequence in which they were collected.

The "Telephonoscope" by 19th century artist Albert Robida

Mechanical television: a Nipkow disk, ca. 1921

To accomplish this seemingly straightforward task, the first attempts at television employed a spinning disk, which was perforated with a spiral sequence of small holes. The earliest description of such a device was proposed by a German, Paul Nipkow, and is usually referred to as a Nipkow disk. As this disk spun, light filtered through the holes and fell upon a photoelectric cell coated with a substance such as selenium, which converted the light into electricity. Bright portions of the image would generate a stronger current than dark portions. The fluctuating current reproduced a semblance of the original image on a similar disk-and-photocell device on the receiving end.

Even at the tender age of thirteen, Philo Farnsworth knew enough to realize that those discs and mirrors could never whirl fast enough to transmit a coherent image. He knew he'd have to find something that could travel at the sort of velocities that Einstein described in other aspects of his theories—in other words, something that could be manipulated near the speed of light itself. He was fairly certain there was a solution to be found in his unseen new friend, the electron.

When a fertile mind has found its way to the threshold of discovery, it thirsts for two things: more information, and somebody to talk to about it. The first, Philo found by securing a part-time job running the school wagon and applying his twenty-five-cents-per-week wages to the purchase of a set of encyclopedias. He also enrolled in a "Radiotrician" correspondence course with the National Radio Institute.

Somebody to talk to showed up in the form of Justin Tolman, an avuncular, bespectacled, middle-aged gentleman who taught senior chemistry at Rigby High School. In the fall of 1920 when Philo enrolled as a freshman, he signed up for every math and science class the school would let him take, but he quickly found the material too elementary for his needs and set his sights on the senior chemistry class. Tolman, when first approached, laughed at the audacity of a freshman wanting to take a senior course. "We just don't allow that sort of thing," the teacher told Philo. "Come back when you're a senior."

Philo wasn't about to wait that long. He went to the principal, then to the school board, to no avail. In the meantime, he learned that his freshman algebra teacher was having problems with his eyes, so he volunteered to serve as his assistant. Before the year was over, Philo was teaching the algebra class himself. At the same time, he was devouring all the material in his freshman science course. At the end of the semester he brought Tolman a note from his science teacher stating that he'd already finished the full year's course. Still, Tolman was reluctant to accept the young student in the senior class.

"Well then," Philo persisted, "if I can't actually take the course, could I just sit in on the classes?"

"Okay," Tolman conceded, "I guess there's no harm in that."

So freshman Philo began sitting in on the senior chemistry class. Within a week he started taking part in the class discussions. When Tolman realized that his prodigy was already at the level of students who had been in the class all year, he offered to tutor him for an hour after class each day to get him entirely caught up. It quickly became apparent to Tolman that he was tutoring the smartest student he would probably ever meet in his life.

In their after-class tutoring sessions, Philo asked Tolman dozens of questions. Tolman didn't know all the answers, but he had a pretty good idea of where to look, and he had lots of books to lend. Invariably, Philo absorbed them and came back for more.

Tolman was amazed at his pupil's grasp of some of the most challenging concepts of the time. He was astounded one day when he ventured past a study hall and lingered in the doorway while Philo stood at the blackboard and delivered a detailed critique of Einstein's theories of relativity. It was, Tolman recounted later, "the most clear and concise description of relativity" he'd ever heard.

A high-school freshman, 1922

Tolman's curiosity about this insatiable student was aroused. He knew there was something driving this thirst for knowledge, but he did not yet know what it was.

The notion of television never stopped tugging at Philo's imagination. In his relentless pursuit of the subject, he learned more about the properties of electrons. He learned how they could be deflected by magnets. He also learned how certain substances could be caused to glow when bombarded by electrons within something called a "cathode ray tube." With those three elements—the electron, magnetic deflection, and the cathode ray tube—he began to believe he would find a solution.

While the great minds of science, financed by the biggest companies in the world, wrestled with 19th century answers to a 20th century problem, the summer of 1921 found Philo T. Farnsworth, age fourteen, strapped to a horse-drawn disc-harrow, cultivating a potato field row by row, turning the soil and dreaming about television to relieve the monotony. As the open summer sun blazed down on him, he stopped for a moment and turned around to survey the afternoon's work. In one vivid moment, everything he had been thinking about and studying synthesized in a novel way, and a daring idea crystallized in this boy's brain. As he surveyed the field he had plowed one row at a time, he suddenly imagined trapping light in an empty jar and transmitting it one line at a time on a magnetically deflected beam of electrons.

This principle still constitutes the heart of modern television. Though the essence of the idea is extraordinarily simple, it had eluded the most prominent scientists of the day. Yet here it had taken root in the mind of a fourteen-year-old farm boy.

It seems quite unlikely that an unknown high school freshman with little education, no money, and no equipment could steal the race for television from the most accomplished engineers and the greatest electrical companies in the world, but with this flash of inspiration, that is precisely what Philo Farnsworth set out to do.

When he told his father what he'd come up with, Lewis cautioned his son not to discuss his idea with anyone. Ideas, he reasoned, are too valuable and fragile, and could be pirated easily. But Philo had to talk to someone. He needed to hear from somebody besides his father that his idea might work.

Late one afternoon in the spring of 1922—on the very last day of the school semester—Justin Tolman finally learned what was driving his young prodigy. After all the other students had left the building, Tolman returned to his classroom and was startled to see a complicated array of electrical diagrams scattered across the blackboard. At the front of the room stood the lanky Philo Farnsworth, chalking in the last figures of the final equation.

"What has this got to do with chemistry?" Tolman asked.

"I've got this idea," Philo calmly replied. "I've got to tell you about it because you're the only person I know who might understand it." The boy paused and took a deep breath. "This is my idea for electronic television."

"Television?" Tolman asked, "What's that?"

The young inventor spent several hours that afternoon with Tolman, elaborating upon his idea. Tolman could not fully understand what the boy proposed to do, nor how he would go about it, though he could grasp the magnitude of the idea. As the lengthening shadows of that Idaho afternoon stretched into dusk, Tolman could do little more than offer vague encouragement, trying to assure Philo that he could do whatever he put his mind to.

Philo had adopted the habit of carrying a small pocket note-book with him wherever he went, so that he could easily jot down the ideas that came to him whenever inspiration struck. As the conversation with Tolman wound down, he pulled out his notebook and drew one more simple sketch of his idea.

"Hang on to this," Philo said, handing the sketch to Tolman, "you never know when it might come in handy." Tolman nod-ded, folded the little piece of paper, and tucked it away inside the pocket of his coat. When their discussion ended, they walked out of the Rigby High School building together and said their good-byes. They would not see each other again for more than thirty years.

At the end of another harvest, in the fall of 1922, Lewis Farns-worth packed up his family and moved again, this time to Provo, Utah. Philo never did graduate from Rigby High School, but nev-ertheless turned his sights on the Brigham Young University, hop-ing to enroll in college level math, science, and physics courses. Unfortunately, since he lacked sufficient funds for the tuition, he chose instead to stay behind in Glenn's Ferry, Idaho while the family moved to Provo.

With the help of his half brother Lewis and a foreman des-perate to find a capable electrician, he found a good-paying job working for the railroad. Philo was barely sixteen years old when he applied for the job, but he didn't blink when the foreman looked him straight in the eye and said, "You're twenty-one, right?"

Using what he'd learned from his correspondence courses with the NRI, he stayed on the job for nearly a year, earning enough money to start classes at BYU when he rejoined his fam-ily in Provo in the fall of 1923. Unfortunately, the authorities at BYU were not sympathetic to his desire to begin studying at the college level. He had neither a high-school diploma, nor suffi-cient English or history. So rather than college level math, science, and physics, Farnsworth was compelled to spend his hard-earned tuition money taking prep-level courses at the BYU High School.

Rigby High School, Rigby, Idaho

Eventually, he was granted "special student" status, and after completing his high school curriculum, was admitted to a higher-level math class. He also gained access to the university's glass lab, where he saw his first vacuum pump, traps, glass arcs, and other tools of the vacuum tube trade. Taking full advantage of the opportunities, he began developing the skills he would need to fabricate his invention—if he ever got the chance.

In Provo, the family took up residence in a large two-story house, renting out the top floor to BYU students. Lewis Farnsworth had moved his family to Provo so the children would have access to better schools, but found that there was not enough work for him in the city. To supplement his income, Lewis resumed taking freighting jobs, often venturing out in harsh conditions in the mountains around Provo. In the late fall, he found steadier work at a resort in Warm Springs, Idaho, using his team and wagon to carry loads for construction projects around the grounds. On one such trip shortly before Christmas, while crossing the mountains in a freezing rain, Lewis contracted pneumonia and barely made it back to Provo with enough strength to fall, sick and exhausted, into his bed.

His children had managed to raise enough money to buy their father a new suit for Christmas, but he never gained enough strength to get out of bed to try it on, or to participate in any other holiday festivities. Shortly after the New Year, gravely ill, Lewis summoned Philo to his bedside.

"Son, I'm leaving you in charge of the family. Take good care of them."

Clutching desperately to his father's hand as he slipped away, Philo fought back the tears and promised, "I will Papa." Lewis Farnsworth was only fifty-eight. Philo was devastated at the loss. He had so many plans. He and his father had been so close. After the funeral, he walked four miles to his father's grave and tried to pull himself together.

Besides losing his closest confidant, Philo—barely sixteen years old—suddenly found himself responsible for the care of two younger sisters, two younger brothers, and his grieving mother, who collapsed into a prolonged depression and took little interest in her surroundings for several months after her husband's passing. The older of the two girls, Agnes, together with a cousin living with the family, took charge of the domestic affairs, cooking and maintaining the family's quarters on the ground floor as well as the boarding house on the second floor. Philo was forced to leave BYU in search of whatever work he could find. The likelihood of developing his idea for television seemed discouragingly remote.

At one point, Philo told a friend that he was thinking about writing up his television ideas and submitting them to *Popular Science*. He thought he might be able to make $100 if he worked it right. But convinced by his friend that publication might not be the most prudent course, Philo instead registered with the University of Utah placement service in hopes of finding work.

Two

The Other Woman

Elma "Pem" Gardner—literally the girl next door

"Gravitation is not responsible for people
falling in love."
—Albert Einstein

In the spring of 1926, George Everson and Leslie Gorrell were driving from Los Angeles to Salt Lake City by way of the Mojave desert when Everson's car, a 1922 Chandler Roadster, burned out a main bearing. Abandoning the car at St. George, Utah, Everson and Gorrell proceeded by bus and train. The mechanic promised to drive the car to them in Salt Lake when he finished the repairs.

Everson was a professional fundraiser on his way to organize a community chest campaign in Salt Lake City. His career had taken him into some of the West Coast's tightest financial circles. As he traveled from city to city organizing for a variety of good causes, he hired local college students to staff his operations. In Salt Lake, he contacted the University of Utah placement service.

One of the applicants was nineteen-year-old Philo T. Farnsworth, present occupation—none.

The preceding two years had not been easy for Farnsworth. When his father died, the aspiring inventor had been forced to curtail his formal education and support his family, developing his television ideas in quiet solitude.

Philo did not abandon his studies altogether, arranging for classes whenever he could find the time between his odd jobs. He even found time for a bit of a social life. After graduating from the BYU High School program, he worked his way into a variety of activities. He went on moonlight hikes and attended school dances. And he played first violin in the school orchestra, where he met and started dating a young classmate named Rae Rust.

Rae's father took a liking to his daughter's suitor, and sensing his rather dire financial straits, offered him a job in forestry near the Four Corners region of Utah. But Philo didn't care much for forestry work. Since his money for BYU tuition had run out, he decided he could get the education he needed if he applied to the Naval Academy.

He scored high on the entrance exam—so high, in fact, that the navy wanted to recruit him for officers school. First he had to make it through boot camp in San Diego, California—in August. The parade drills proved so exhausting that many of the recruits just fell down in their tracks, and Philo often wanted to drop, too. He managed to stay standing through the rigors of boot camp, but he knew that military life was not for him.

A navy chaplain who befriended Farnsworth was inclined to agree. Philo confided his big dreams to the chaplain, who cautioned him that the navy could appropriate his invention if anything came of it while he was enlisted. The chaplain was also sympathetic to Farnsworth's plight as the primary breadwinner of his family, and arranged for a "humanitarian discharge."

This brief stint in the military was not without a lasting impact, though. Discovering that his unusual first name would provoke ribbing from his fellow Seabees—who were inclined to confuse "L" with "D" and call him "Fido," Farnsworth dropped the "O" from his name, and forever after was known simply as "Phil."[5]

He had managed to save a little money while in the navy—just enough to return to Provo to resume classes at BYU—and pick up where he'd left off with his girlfriend, Rae. Meanwhile, Phil's sister Agnes had become close friends with Elma Gardner, known to her friends and family as Pem. The Farnsworth and Gardner families shared some history—fathers Lewis and Bernard had been friends during their freighting days in the mountains in the months before Lewis died. Though the Gardners were living "out in the sticks," Pem and Agnes were classmates, taking as many classes together as they could. And Agnes never missed an opportunity to brag about her big brother in the navy.

Serena Farnsworth, Agnes and Phil's widowed mother, was well recovered from the depression that followed her husband's death and was working out of the home. She often left a pot of beans on top of the wood-burning stove so the kids could have a hot lunch when they came home from school. Agnes frequently invited Pem home for lunch.

Agnes may have had a bit of an ulterior motive. Leaving Pem alone in the kitchen one such occasion, Agnes abruptly disappeared into another part of the house. While Pem was setting the table, Phil came through the kitchen door, taking her completely by surprise. Pem ran out of the kitchen to find Agnes, telling her, "There's a strange man in the kitchen!"

Agnes returned to the kitchen and introduced her friend to her brother, and the three of them sat down to a lunch of hot baked beans. Recounting the moment years later, Pem recalled, "I don't think either of us was awfully impressed with the other. He had a very intelligent look about him, but I knew he was a college student and I was a high school student, so he wasn't going to be interested in me. I was rather quiet. Agnes and Phil did all the talking, and when lunch was finished we went back to school and that was that."

Although Pem didn't think much of the encounter at first, she did find herself drawn into Phil's active social circle. One idea

Phil came up with was for something called a "radio party." Having found work at a furniture store in Provo that sold radios—his job was delivering the consoles and setting up the antennas—Phil arranged to borrow a top-of-the-line Magnavox console from his employer. He then mailed requests to radio stations in St. Louis, Chicago, Los Angeles, Denver, and other cities that delivered strong signals to the Salt Lake area, asking that certain tunes be played at certain times during the evening. Throughout the evening of the gathering, he twisted the dial and the guests all heard the announcer crackle in with "This request goes out to Philo T. Farnsworth and party in Provo, Utah."

Phil invited his current flame, Rae Rust, but he also invited Pem, Pem's brother Cliff, and other friends from the Provo area. Pem was rather confused by the invitation. "I didn't know if he was going to pick me up or what. He told me that he was bringing Rae to the party, so he's sending somebody for me. I knew this other fellow from high school but I didn't like him too well, I thought he was a rather fresh guy. At the party, he and a couple of the other boys went outside to the car to get a little …refreshment. He got a little too…exuberant…and when he came back into the party he got a little fresh with me, and I slapped his face."

Pem told Phil, "I'm not going home with this guy" so Phil offered to take her home, but had to take Rae home first. "I'll have a talk with this guy," Phil offered, "and I'm sure he'll be a gentleman and take you home." Pem made it home that night without further incident, but couldn't help a healthy laugh when her escort slipped in the mud outside the Gardner home and fell flat on his face.

Meanwhile, Phil had some relationship issues of his own to contend with. As he drove Rae home that night, she expressed her reservations about their future, telling Phil that she couldn't be interested in anybody with as little ambition as he seemed to have. Perhaps his timing was bad, but he thought he might be able to dispel Rae's doubts by telling her about his ideas for a device that could send "pictures through the air." But it all sounded entirely too fantastic and whimsical to Rae, who was looking for something more immediately substantial in the way of a suitor and eventual husband. She could not fathom how something as far out as this "telegraph pictures" thing would be their path to domestic tranquility. Nor did it matter to her that Phil had carried her picture with him throughout his stint in the navy. She was giving up on him because he had no ambition, or, at least in her eyes, the wrong kind of ambition. A while later, perhaps sensing the error in her judgment, Rae tried to win Phil back by writing him a touching poem, but by then his attention had turned elsewhere.

Phil arranged to take another semester at BYU in the spring of 1925, taking classes in trigonometry and differential calculus, among other things. He endeared himself to his professors, and spent many hours in the glass-blowing lab learning still more about the skills required to assemble vacuum tubes.

Once again, Phil's family moved, this time from their two-story residence and boarding house into a more affordable duplex in Provo. When the other side of the duplex became vacant, the Gardner family moved in. The move seemed a natural, as the two families had grown close. Cliff, the oldest of the Gardner boys, was nearly the same age as Phil, and since the two boys shared a common interest in radio and things electrical, they became good friends. Phil's brother Carl and Cliff's brother Art were also the same age. And Agnes and Pem had been classmates and friends for a couple of years.

Pem was the one to ask Phil out on their first real date. Pem's sister, Verona, had a boyfriend who had a car, something of a rarity among kids in those days. Verona wanted to go on a picnic to Warm Springs in Provo Canyon and suggested that Pem invite Phil and Agnes to join them.

"Not on your life," Pem replied. "I'm not going to ask a boy to go with me. I never have and I never will."

Despite her objections, Pem's sister prevailed, and Pem went over to the Farnsworth's side of the duplex, first to find Agnes, and together they found Phil and invited him to join the party.

"Sorry," Phil said, somewhat sheepishly, "I'm boning up for an exam on Monday." He must have seen the annoyance in Pem's eyes. After all, this was the first time she'd ever asked a boy to go out with her, and here he was turning her down! Sensing her disappointment, Phil relented and agreed to join the group for their picnic. Later Pem found out the real reason he'd declined at first was that he was flat broke.

Fortunately, Cliff Gardner had just returned from a job he'd had working in the coal mines at Castle Gate, and he was loaded. So off they all went to the Sarasota Warm Springs, a fashionable resort with pools, a huge dance floor, and a jukebox. Cliff fed nickels into the machine and they all had a great time dancing the Lindy and the Charleston, followed by a swim. They sang Hit Parade songs—"Margie," "Sheik of Araby," "All Alone," and "Somebody Stole My Gal." And somewhere in the course of the day, Phil and Pem began to sense something special between them.

Music played an important part in the young couple's lives. Phil played violin, and when he found out that Cliff played trombone and Pem played piano, he'd bring his violin to the Gardner's side of the duplex and the house would ring with impromptu jam sessions. One day after such a session, the three went for a walk. Suddenly Phil got a crazy notion, and picked Pem up, and set her on top of a hedge.

"I'll take you down when you kiss me," Phil dared.

"You get me down off of here or I'll kick you," Pem replied, recalling, "Phil got his face slapped for that, but he got his kiss later on."

With jobs still hard to find in Provo, Phil divided his time between working for the radio store in Salt Lake City and spending weekends with Pem in Provo.

As if to underscore Phil's growing affection for Pem, he took a step like the one that had led to the breakup with his previous girlfriend. Though he feared a similar result might ensue, Phil shared his secrets with Pem, especially his dream to be an inventor, and to launch his career with something called "television."

"I thought it was just too far out," Pem recalled, but her doubts did nothing to diminish her deepening devotion. Taking Phil's hand and gazing into his deep blue eyes she confessed, "If you told me you could fly to the moon, I'd believe it."

Philo T. Farnsworth in 1926

By Christmas of 1925 Pem and Phil had something of an understanding, and were dating each other exclusively. As they returned home with a group of their friends from a holiday gathering in nearby Salem, their car ran out of gas, and while Cliff went off to find a filling station, Phil took the initiative. "I think we're meant for each other," he said. "What do you think?" Pem shared Phil's feelings, but they both felt the need for more education, so any talk of marriage was going to be "some time off."

Meanwhile, an equally strong bond formed between Phil and Cliff around their mutual interest in all things electrical. In the spring of 1926 the two boys ventured to Salt Lake City to start their own business of installing and repairing radios. Unfortunately, competition in the local radio business was tough, and Phil's first attempt at running his own business did not fare well. Hoping to find more work to continue providing for his family back in Provo, Phil turned to the University of Utah placement service.

When Phil Farnsworth showed up in George Everson's office for an interview, Everson was looking to fill a number of small jobs conducting community surveys. As usual, Phil had better ideas. He volunteered himself right away as the Survey Manager and assured Everson that he was so familiar with the territory that he was indispensable. Everson, who "possessed the instincts of a gold rush gambler with his nose too close to the ground," hired the eager and confident young applicant immediately. Everson always knew when he picked up a good scent, even if he wasn't quite sure what he was smelling.

Phil's first responsibility was to complete the job of hiring the campaign staff. Demonstrating the devotion to family and friends that would soon become a trademark of his career, his first recruits were Cliff and Pem Gardner and his sister Agnes. Until then Pem and Phil had spent time together in Salt Lake City only on occasional weekends when Pem's mother, whose health was failing, would permit her to go. When Pem's mother died in the winter of 1926, the seventeen-year-old became responsible for the rest of the Gardner brood and her visits had become even less frequent. But with the prospect of a good job, Pem left Provo and took her own room in the boarding house where Cliff and Phil shared one.

George Everson got a better idea of Phil's capabilities a short time later, when word came from St. George that his Chandler Roadster, which he'd left behind enroute to Salt Lake, had only made it as far as Beaver City before the bearing burned out again. George instructed the mechanic to make the necessary repairs, and Phil persuaded George to let him make the trip to fetch the car.

Two days later, George took a collect call from Provo. "Where are you?" George asked impatiently.

"I'm in Provo," Phil replied. "The bearing burned out again. When I couldn't get it fixed, I just pulled the piston out and have limped in here on five cylinders. I should be in Salt Lake by late tonight."

The next morning, Phil impressed George as "knowing more about the car than the man who made it." The event served to bring George and Phil closer together, and gave George new respect and admiration for his young charge's abilities.[6]

After a few weeks, as the survey was winding down, George discovered that an important mailing had not gone out on time, so the entire staff stayed after dinner to help Phil finish the job. Afterward, George, his colleague Les Gorrell, Cliff, and Phil paused for a casual bull session. Having gotten to know one another by then, George was curious about what might be lurking behind the high forehead and sparkling blue eyes of his Survey Manager, and asked Phil if he planned to go back to school.

"No," Phil replied. "I can't afford it. I've been trying to find a way to finance an invention of mine but it's pretty tough." Phil explained that he'd been thinking about it for nearly five years, though, and was quite sure it would work. "Unfortunately, the only way I can prove it is by doing it myself; but I don't have any money. I've even been thinking about writing it up as an article for one of the popular science magazines—I might be able to get a hundred dollars for it."

"What's your idea?" Les Gorrell asked.

Phil paused before he answered. "It's a television system."

George, who had never heard the term before, asked curiously, "Tell-a-who?"

George and Les listened intently as Phil described the scheme that had captivated his imagination for the past five years. George, who had little knowledge of engineering or science, commented, "That's really quite a pipe dream you've got there." But Les Gorrell, who was a graduate of Stanford University with some knowledge of engineering, was intrigued. Phil went home shortly after, frustrated that he hadn't generated any interest in his idea. However, Les continued the discussion privately with George.

"I'd take another look at that," Les told George, "He might have something." The next evening, they invited Phil back for a more serious discussion.

Greatly encouraged that they wanted to continue the discussion, Phil was transformed with confidence. With a receptive audience for his ideas, his manner changed from what George described later as that "of an office clerk too closely confined to his work." As Phil spoke that night, a special power came to him. His bright blue eyes darkened with intensity as he spoke of

The partnership of Everson, Farnsworth and Gorrell

the ideas that had occupied his brain for the last five years. His speech found new eloquence as he became charged with the energy of his own genius.

George remained the skeptic. He suggested that GE or AT&T must have already accomplished what Phil proposed. Phil countered with a detailed treatment of just what was going on around the world. He described the mad professors and their wonderful, spinning wheels. "They're all barking up the wrong tree," he said.

When Phil had answered all of George and Les's questions, George finally asked how much it might cost to build a working model of the machine he had described. Taking a shot in the dark, Phil said it might cost about $5,000. "Well," George said, "your guess is as good as any. I surely have no idea what is involved. But I have about $6,000 in a special account in San Francisco. I've been saving it with the idea that I'd take a long shot on something and maybe make a killing. This is about as wild a gamble as I can imagine. I'll put the $6,000 up to work this thing out. If we win, it will be fine, but if we lose, I won't squawk."

In short order, the association of Everson, Farnsworth and Gorrell was established. Farnsworth insisted on nominal control of the association, and for the contribution of his invaluable genius he was awarded half the equity in the company. In exchange for raising the money, Everson and Gorrell would split the remaining half.

With the partnership formed, George suggested that Phil consider setting up his operation in Los Angeles. Phil agreed that was a good idea—the resources of a vast metropolis like L.A. would be much more suited to finding and fabricating parts for his exotic apparatus. There was just one more detail to square away.

"I'll have to get married before I can go," Phil said.

This particular revelation took George by surprise, but he wisely concluded that it would be far better for Phil to get married than to have half his mind on a girl in Utah and half his mind on his work on the West Coast.

That evening, Phil called Pem, who had returned to Provo to live with her family. "Pem, I've got backing for television, can you marry me in three days?"

The families—especially Phil's mother and Pem's father—were startled by the sudden change in fortune and skeptical that a marriage conceived in such haste could survive. Phil was nineteen at the time; Pem was eighteen. But Phil made a strong case.

"I can help the whole family," Phil told his mother. "I can help *both* families better if we do this than if I stay here." Turning to Pem's father he added, "This is my big chance, and I really feel that Pem should go with me."

Mr. Gardner thought about it for a moment, but that was as long as it took to add his blessing to the unfolding hand of destiny. "I'll find work here in Provo and take care of the family," he told his daughter and future son-in-law. "You kids go ahead and make the best of it."

Two days later, the wedding was performed by a Mormon bishop at the duplex in Provo. After a brief reception, Phil and Pem drove George's Chandler back to Salt Lake City and checked themselves into a nice hotel. While Pem prepared herself for their private celebration, Phil kissed his bride—and went off to find George, to arrange a little pocket money for the trip to Los Angeles. Neither George nor Phil must have realized the significance of this particular evening, because their nocturnal visit turned into a long discussion of the future of television. Meanwhile, Pem's emotions swung between disappointment at seeing her wedding night diminished and fear that something simply awful had happened to her new husband. As the hours slipped by, she fell asleep.

When Phil finally returned and awakened her, he could see the anxiety on her face, and tried to restore her mood with a joke.

"Pem," he offered in jest, "there is another woman in my life . . . ."

When the shock passed and Pem began to catch on to the joke, Phil finished his sentence: ". . . and her name is television."

THREE

The Daring of This Boy's Mind!

Newlyweds Phil and Pem in Santa Monica

"Be bold and mighty forces will come to your aid."
—Johann Wolfgang von Goethe

The newlyweds consummated their marriage on a Pullman train from Salt Lake City to L.A. This was the first time Pem had been out of Utah, so of course her father had dutifully admonished her about the sins of the Big City. On arrival, Pem recalls, she was awestruck. "I just couldn't believe any place could be so big!"

Their honeymoon consisted of a Sunday afternoon spent strolling the beach in Santa Monica—the first time Pem had seen the ocean. They rode the roller coaster on the Santa Monica Pier and had a souvenir picture taken in front of a mock-up of the train that had brought them to the coast. And then it was time to get to work.

When Phil had dropped in on George on his wedding night, he had received a check and the name of a colleague in Los Angeles who would cash it for him. So after the newlyweds secured temporary quarters in a little efficiency apartment on Lankershim Boulevard, Phil located Harry Cartlidge on the Universal Studios lot, where he was running a hospital fundraising campaign. Harry took Phil to lunch at the studio commissary. There they dined with the likes of George O'Brien and Mary Livingston, and other actors and actresses wearing blue lips and green hair—makeup suited for the peculiar sensitivities of the black-and-white films used in those days.

Meanwhile, Pem began to prepare her first dinner as a newlywed. She soaked some beans for a couple of hours, "but these beans just stayed hard all day." When Phil got home, he found his bride near tears and offered to take her out for dinner. At the restaurant, they picked up a newspaper and came across an ad for an apartment at 1339 North New Hampshire Ave. After dinner they boarded a streetcar to a quiet neighborhood just off Sunset Boulevard, and Mrs. Knapp, the landlady, showed them a tidy one-bedroom apartment on the ground floor of a two-story four-plex. Looking around the place, Pem saw enough space in the living room for a piano, and Phil saw enough space in the dining room for a workbench, so they agreed to take the apartment on the spot, for $45 a month. And so the future parents of television established their first residence together, not far from the heart of glamorous, Roaring Twenties Hollywood.

Turning to the task at hand, Phil realized the assignment he had given himself was doubly difficult. Before he could build the marvelous machine he had sketched for his high school teacher four years earlier, he had to design and build many of the tools necessary to proceed. It was not as though he could run out to a TV parts store and pick up whatever he needed. This was new territory, and virtually everything had to be made from scratch. Pem described the process as like "building a whole car from parts, just to go to the store for groceries."

Along the way, Phil began to acquire still more specialized knowledge: electrochemistry, radio electronics, optics, and the ancient art of glass blowing. Although the glass-blowers he met said that the tube he wanted was impossible to make, Phil, true to type, ignored their opinions and proceeded to do what had to be done.

He knew instinctively that the $6,000 George and Les Gorrell had staked him to was not going to be enough to construct a working model of electronic television, let alone demonstrate anything even remotely commercial. Rather, the Los Angeles effort would have to be directed more toward demonstrating his own abilities, to build a basic circuit or two, to show George and Les that he could really "do this thing" that he had promised he could do. He decided to set his sights on building a deflection coil, not only to demonstrate that electrons would do his bidding within a magnetic field, but also to prove that he was capable of making them do so.

George similarly suspected that his $6,000 investment would not go very far, and had his own ideas of what would come next. What he hoped to do was simply to get Phil on his feet, and to gain enough of an understanding of his own so he could make a viable presentation of Phil's ideas to other, more substantial backers. Toward that goal, he did all he could to help Phil with his various projects.

George and Les were in and out of Los Angeles all summer, helping Phil any way they could when they weren't off on another fundraising campaign. They chauffeured Phil around the city, looking for equipment such as meters, power sources, and a drafting table—where Les made the first drawings of Phil's invention and taught Pem the art of technical drafting so that she could continue the work when Les was off on other business.

One Saturday afternoon, Phil needed help with winding the wire coils to serve as his first magnetic deflection yoke. George was the most unlikely of choices for such a task. He was fastidious in his attire, whereas winding coils was a messy job that involved insulating each layer of wire with a coating of shellac. Nevertheless, George rolled up his sleeves and set about winding coils in the garage behind the New Hampshire Avenue apartment.

It must have seemed a bit curious—if not downright bizarre—all this activity in and out of this little apartment, especially considering that it was all taking place during that infamous time in American history known as "Prohibition."

At the end of each day's foraging around the city, George, Les, and Phil would return in George's sporty Chandler and disappear with mysterious bundles into the little apartment. When they weren't out searching, Phil was conducting experiments with crystals and optics, using a focused light source that required the window shades to be drawn during the day. And George was often showing up with friends and colleagues to show them what was going on. Guests would arrive, the shades would be drawn, and they would not be reopened until just before the guests were ready to leave.

On this particular afternoon, George, a total stranger to the neighborhood, was stationed in the backyard winding copper wire around a cardboard tube. Up to his rolled-up elbows in shellac, he came upon a problem, and came in through the back door to find Phil with a question.

At that very moment, Pem was answering a loud knock at the front door and found her porch "filled with two of the biggest policemen I ever saw...standing so that the whole door frame was filled with police!"

The sergeant of the squad confronted the startled Mrs. Farnsworth. "Ma'am, we've had a report that you're operating a still at this address." Pem was taken aback at the charge, and said, "Wait a second. I'll get my husband." By then Phil was already right behind his wife, and didn't lose a beat, inviting the officers in to show them what he was doing, just as if they were some of George's friends.

Meanwhile, George had just come into the apartment through the back door. Hearing the commotion, he decided this was no place for a gentleman of his stature to be found. So he turned around and headed back out the back door—only to be confronted by another pair of oversized policemen who stopped George in his tracks.

"Oh no ya don't, buddy," one of the cops said, as he ushered the flustered, sputtering George back inside the apartment.

With all the suspects assembled in the living room, the police proceeded to have a look around the apartment. They found nothing alcoholic, but seeing a Boris Karloff collection of strange equipment, the sergeant began to wonder if they had stumbled onto something even more sinister than an illegal still.

An early magnetic coil

"Okay, Mister," the sergeant wanted to know, "what is all this stuff?"

Phil looked at the strange gear he had collected in his dining room, stared the sergeant straight in the eye, and answered, "This is my idea for electronic television."

The sergeant shook his head, took another look around, and said, "Tell-a-what?"

Phil proceeded as best he could to unravel the mystery, explaining that he was working on a machine that could make movies work like radio. That seemed sufficient explanation for the Hollywood police.

"Well then, we're sorry," the sergeant said, "I guess there's been some mistake," and with that the squad took their leave of Farnsworth and Company. It was a good thing they didn't search more thoroughly, or they might have found the bottle of gin that Les Gorrell kept hidden "for special occasions."

However, that was not the end of the episode. Somehow, the police learned that the Farnsworth's next-door neighbor was none other than Hazel Keener, a starlet who had most recently played opposite Harold Lloyd in *The Freshman*. Later that afternoon, Pem was talking with Hazel's mother over the fence, when Mrs. Keener asked about her visitors.

"Yes," Pem said, "we had a visit from the police. They apparently thought we were operating a still."

"Oh, so that's what they were looking for," Mrs. Keener said. "They came over to our house, too. Hazel had just gotten home from the studio, and as usual had just gotten comfortable in a robe when the police came in. Once they got sight of Hazel, well, they just took their sweet time looking everywhere, including in the closets and under the bed!"

With their Keystone Cops adventure behind them, Phil and his makeshift team returned to their work. About a week later, Phil was ready to test his circuits. As George Everson later recounted:

"I had just bought fifty-four dollars worth of tubes, which were put in a chassis. The motor generator was turned on, and we all stood around expectantly, hoping to see the results of a beam of electrons deflected by magnetic coils. Farnsworth had not guarded against the surge, as the motor generator was started, and the whole batch was burned out in a split second."[7]

Phil, Pem, and George all watched, mortified, as Phil's first experiment with electronic television snapped, crackled and popped into a plume of smoke. Their noses filled with the acrid aroma of burning wires and coils. Within a few seconds, it was all over. When the smoke cleared, Phil realized he should have been more cautious. With a note of panic in his voice, he said, "I'm sorry, George, but that's all I have to show you for your investment."

"It's not the end of the world," Les Gorrell chimed. "After all, we still have Phil's ideas."[8]

Les had the right idea and George knew it. In fact, he had suspected all along that little would actually result from this modest investment and makeshift setup. In Phil Farnsworth, George knew he had found somebody with a special gift, and he wasn't about to let a tangle of molten tubes and wires stand in the way of Phil's electronic rainbow—or the fabulous pot of gold George was certain he would find at the end of it. In fact, he was already planning his next moves, and thinking about how he was going to raise more money for this dream.

For George, it was one thing to put his own "mad money" into a venture as far-fetched as Phil's invention. Bringing other investors into the project was something else altogether, something that could put his reputation in the financial community at risk. He could not afford to jeopardize his standing by acting prematurely. Lacking the technical background to make a sound judgment on these matters, George thought it would be prudent to call upon a more reliable source.

George and Phil were on the same wavelength now, although Phil had slightly different concerns. Before he would feel comfortable discussing his ideas with any more potential investors, Phil felt it imperative to seek the protection of the United States Patent Office, to get his ideas in writing with the office that governed such affairs.

Fortunately, Les Gorrell, the former graduate student in engineering, recalled a couple of classmates, Leonard and Richard Lyon, who had since become highly regarded patent attorneys in Los Angeles. George and Phil both agreed with Les that Lyon and Lyon should be consulted before going any further, and before seeking any additional funds for Phil's experiments. In short order, an appointment was made and the partnership of Everson, Farnsworth and Gorrell found itself in the wood-paneled offices of Lyon and Lyon, patent attorneys.

Once again, when called upon to describe his invention, Phil seemed almost possessed by mystical forces. He projected not only his ideas, but himself, conveying an unmistakable sense that he knew precisely what needed to be done and that he could do it. The Lyon brothers listened intently, and peppered Phil with detailed questions that he answered with poise and knowledge well beyond his years and his level of formal education. Finally, Leonard Lyon expressed what was going on his own mind:

"If you have what you think you have, then you've got the world by the tail. If not, the sooner you find out, the better."

The Lyons were fairly certain that Farnsworth's ideas were sound, but they, too, wanted another opinion. Arrangements were made for Phil and Leonard to meet with Dr. Mott Smith of Cal Tech, who would pass further judgment on the merits of the proposal.

When Dr. Smith arrived for the session a week later, he paid in advance for only one hour of parking, fully expecting to dismiss the scheme and leave. However, the meeting wore well into the afternoon. As Pem recounted later, the atmosphere that afternoon was "electric." Phil knew exactly what he had to do that day, and once again rose to the occasion. Throughout the meeting, Leonard Lyon kept getting up from his seat, "pacing the floor like Felix the Cat," and commenting audibly on the enormity of what he was hearing:

"It's preposterous! Impossible...It's just amazing...the daring of this boy's mind!"[9]

Once the Lyons and their expert were satisfied that Farnsworth had answered all their questions, George needed to know just three things:

"First," he asked, "is this thing scientifically sound?"

Dr. Smith answered, a bit bemused: "Yes."

"Is it original?" George continued.

"I'm pretty well acquainted with recent electronic developments," Dr. Smith replied. "I know of no other work that is being carried out along similar lines."

Finally George wanted to know: "Is this thing feasible? Can it be worked out to make a practical operating unit?"

In his answer, Dr. Smith could only imagine the road that lay ahead: "You will encounter great difficulty in doing it, but I see no insuperable obstacles at this time."[10]

That was all George needed to hear.

FOUR

The Damn Thing Works

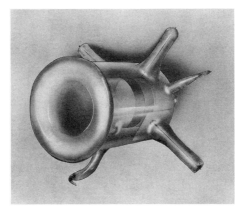

The very first Image Dissector tube, ca. 1927

"Every truth passes through three stages before it is recognized. In the first it is ridiculed, in the second it is violently opposed, in the third it is regarded as self-evident."

—Arthur Schopenhauer

With patent attorneys on board, the process of protecting the inventor's ideas was underway, and George could begin the quest for more substantial funding for the experiments.

Phil told his benefactor that he thought $1,000 a month for twelve months would be enough to come up with a working model of his television. Observing that Phil had a knack for underestimating the financial needs involved, George set about to raise the sum of $25,000.

He called on all his contacts in the world of high finance to find the individuals who might have sufficient capital to back the project. In the process he encountered a colorful cross-section of California's wealthiest society—and was turned down for the strangest reasons. It seemed that every rich man already had a special interest that absorbed his "extra" money. One industrialist, obsessed by color photography, expressed interest if the television were in color instead of black and white. Another was interested only if the idea had some application to bacteriology. None of these first contacts could see beyond their own personal interests and recognize the tremendous value in George and Phil's proposition.

Late in August 1926, George Everson appeared in the offices of the Crocker National Bank in San Francisco looking for one Jess McCargar. McCargar, whom George had met some years earlier on a Community Chest campaign, was an officer of the Crocker Bank. But McCargar was on vacation and would not return for some time. Another bank officer, James J. Fagan—known affectionately as "Daddy" Fagan—observed George's disappointment and asked if he could help.

"I don't think it is anything that would interest you in the least," George replied. "It's not an investment, it's not even a speculation. It is wildcatting, and very wildcatting at that."[11]

George's response served only to intrigue Mr. Fagan, who at the time was considered the most conservative banker on the West Coast. He was a wizened, tobacco-chewing veteran of the California Gold Rush. His steely judgment on investment matters had earned him the reputation as a "cold-hearted, glassy-eyed guardian of the moneybags." Whatever the reasons for his interest, Daddy Fagan prevailed upon George to explain why he had come looking for McCargar.

With his crusty banker's sixth sense, Fagan listened as George Everson described the genius he had stumbled across. When George finished explaining Farnsworth's ideas, the Crocker Bank officer drummed his fingers together and spat a wad of tobacco at the gold cuspidor in the corner of the room: "Well, that's a damn fool idea, but somebody ought to put money into it," Fagan said, adding, "someone who can afford to lose it."

Two days later, W.W. Crocker himself suggested in the strongest terms that George summon his young protégé to San

Francisco to meet Roy Bishop, a successful capitalist and engineer of some standing.

When Farnsworth arrived by train in San Francisco, he looked every bit the part of the struggling inventor: frumpy, frayed, and preoccupied. So George took his protégé shopping, outfitting him in a new powder-blue suit, hat, and haberdashery, and together the two set off to meet Roy Bishop for lunch.

Bishop listened intently as Farnsworth described his idea. As the conversation wound down, however, Bishop seemed to be somewhat reluctant. "I am convinced that the idea is sound," he told Phil, "but I doubt your ability to work it out commercially."

Hearing Bishop's reluctance and sensing the negative drift of the conversation, Farnsworth prepared to play his hand. For the moment he was holding all the cards. He knew what needed to be done to create television and he was confident he could do it. But he could not afford to be involved with people who did not have equal confidence. Putting his papers in his briefcase and rising from his chair, Farnsworth courteously thanked Bishop for his kindness and time. As he turned toward the door he said, "I am sorry that you are unable to see the possibilities that I see for this invention."

Stunned, George quickly gathered up his things and caught up with Phil at the door. As they were about to leave, Bishop said, "Wait a minute!"

The strategy had worked; Bishop the engineer demanded only one final stipulation. He wanted to consult with another "hardboiled" engineer, a man named Harlan Honn. "If you can convince him that your proposition is sound, then I think we can find a way of backing you."

Honn was summoned and arrived in less than half an hour. He looked over the specifications, asked some questions, and turned to George with a simple pronouncement: "Why, sure this system will work. I think very well of it."

Bishop reported all his findings to the people at Crocker Bank, and the matter was held in abeyance until Jess McCargar returned from vacation.

The meeting with McCargar and the other principals of Crocker Bank took place in the Board Room, part of the opulent inner sanctum that George sarcastically called "The Throne Room."

The circumstances of the meeting were unlikely at best, for this was a place where many requests were entertained but few were granted. What could this unproven, self-educated, twenty-year-old farm boy possibly say to these crusty robber barons that would compel them to cut loose enough cash to start something as far-fetched as television?

While the bankers huddled in a corner of the Board Room talking the matter over, George and Phil sat on a marble bench waiting to be called. Within a few minutes, McCargar came over and put his arm around George, saying, "I think we are going to back you, boys," and ushered Phil and George into the oak and marble-laden Board Room.

We can only imagine the atmosphere that must have prevailed that day, what cosmic forces had to converge on this room for this moment to happen; for happen it did: Philo T. Farnsworth had swept in with another dazzling display of his unbridled genius, and before the session ended he had been staked to the then-substantial sum of $25,000. He was given the use of a loft in San Francisco where he could set up a laboratory—and one year to produce some kind of television picture.

Roy Bishop underscored the unusual nature of the event when he turned to Farnsworth with admiration in his voice and said, "Young man, you are the first person who has ever gotten anything out of this room without putting up something in return." Then Bishop addressed the rest of the group and delivered an ironic benediction: "We're backing nothing here but the ideas in this boy's mind. Believe me, we're going to treat him like a race horse."

The terms of the arrangement with Bishop and the Crocker group were not ideal from Farnsworth's perspective. The original partnership of Everson, Farnsworth, and Gorrell was dissolved and replaced with a new enterprise, called Crocker Research Laboratories. More important, with the infusion of new capital,

Farnsworth could no longer retain his original fifty-percent stake. He was still the largest individual shareholder, but his share was substantially reduced, making him a minority partner for the first time. While he may have objected to this turn of events on principle, it was clear to him that a minority stake in a going enterprise—a real chance to produce his inventions—was infinitely preferable to a majority stake in nothing.

Knowing little of the workings of high finance, he trusted that George would protect his interests. As Phil told Pem, he was confident that "everything would work out all right." All the papers were signed, with George acting as Phil's guardian, for the young genius had just turned twenty, and was still underage.

Phil returned quickly to L.A. in George's Chandler Roadster to gather up his bride and his little dining-room laboratory. Pem could tell the moment he stepped through the door that things had gone well in San Francisco. With his classy new suit he seemed to radiate success. Phil swept Pem off her feet and they danced around their little living room while he told her of the exciting things in store for them.

Less than four months after their arrival in Los Angeles, in September 1926, the pace of life was quickening again as Phil and Pem packed all their belongings into the Chandler and headed north, singing "Rose Colored Glasses" while they cruised past the cliffs of Big Sur.

Cliff Gardner, Pem's brother and Phil's best friend, was folding cardboard into boxes on an assembly line in Oregon when a telegram arrived. The cryptic message mentioned that Phil had found financial support, and that there was a job for Cliff, but the remaining instructions were a little fuzzy. Nevertheless, Cliff finished folding his last box and walked out to board a train for San Francisco.

Following Phil's instructions as well as he could, Cliff waited at noon for several days for Phil and Pem to meet him near a corner specified in the telegram. He found a boarding house near the designated intersection. Five days after Cliff's arrival in

San Francisco, Phil and Pem found him at the designated intersection, hungry, nearly broke, and happy as hell to see them.

The reunited trio went off to find 202 Green Street, the empty loft that the Crocker group had provided for Farnsworth's new laboratory. What they found was a fairly unremarkable two-story brick edifice at the foot of Telegraph Hill, with their assigned quarters on the second floor above a garage. On the first floor, beneath the sign that said "Crocker Laboratories," they were greeted by Harlan Honn—the engineer who had endorsed Phil's ideas for Roy Bishop—who was developing a refrigeration system as part of another Bishop/Crocker enterprise. At the top of the stairs they found a completely empty loft, with lots of windows and a skylight, but not much else—no benches, no partitions, no equipment of any kind. But that meant they could lay out the space any way they liked. Phil knew the moment he set foot in the place that he had arrived at the birthplace of television.

Pem wasn't too keen on living in the middle of the city, so that night—with the infamous Dempsey–Tunney "long count" prizefight playing on a radio in the background—Phil, Pem, and Cliff took the ferry across the bay to Berkeley to find a home.

"The world was our oyster," Pem would later say of that day. "The heady perfume of the beautiful day and our light hearts all came together in a fairyland world. We were all singing a lot, and between songs, Phil would talk about how the world was going to be for us. He had a lot of confidence in himself, and with the money he now had, with the chance he was getting, he just couldn't fail. He just knew that television was going to make a millionaire out of him."

They headed off to Berkeley, hoping to find a place to live near the campus of the University of California. Turning up Derby Street, they spied a "for rent" sign, and the landlady, Mrs. Wilbur, showed them a nicely furnished two-bedroom apartment "with dishes and linen and everything."

Phil put on the charm with Mrs. Wilbur, telling her, "I'm the inventor of television. This is my wife and this is my assistant. We'll be reliable tenants, and we won't be throwing any wild parties."

202 Green Street, at the foot of Telegraph Hill in San Francisco

Mrs. Wilbur didn't quite catch the part about television, but her proximity to the college campus must have made the idea of renting to a nice, stable married couple appealing.

Now that the Farnsworths had a new home things were truly falling into place.

That night, Phil and Cliff sat down and made a rough sketch of how they would lay out their laboratory at 202 Green. They decided to divide the loft in half: on one side, they would build a tube lab, with a glass lathe and torches, a spot welder, and a vacuum pump table. A small office for Phil, a drafting table for Pem, and workbenches for the other electronics would fit nicely into the other half of the loft. Cliff volunteered for a critical job—to serve as resident glass-blower and tube-builder. His training for the job amounted to a high school diploma, boldness comparable to Phil's, and absolutely no previous knowledge of the subject—a seeming liability that Cliff would quickly turn into an asset. Since Cliff had little idea of what the experts would have insisted "could not be done," there was nothing to prevent him from going ahead and doing it.

At last, Phil could begin fabricating the device that had first appeared in his mind's eye in the middle of a potato field in Idaho five years earlier. Showing his knack for coming up with simple, descriptive names for his inventions, he elected to call his invention the "Image Dissector," because it would dissect an image into individual elements of brightness and darkness, and convert those elements—one thin line at a time—into a pulsating electrical current that could be transmitted through the air.

For the receiving end of his system, Phil started with a standard Erlenmeyer flask—like he had used in his high school chemistry class—for the first picture tube, which he dubbed the "Image Oscillite."

When Farnsworth had finalized the plans for his television system and drawn detailed diagrams, he filed for his first patent. The application was submitted on January 7, 1927. Among inventors, a device is considered invented at the very first "moment of conception." But there was no way to document the date when young Philo had first conceived the Image Dissector among the furrows in that Idaho field. Thus, the date the patent was filed serves as the first official "disclosure" to the Patent Office, and so might be considered the textbook date that television was invented. Nevertheless, the patents could not be officially granted until the device had been proven to work, or "reduced to practice" in patent parlance. Phil and Cliff had no way of knowing how long that would take. They just knew they had promised the financiers that it would done within twelve months.

The first few months in San Francisco were a heady and romantic time for the newlywed Farnsworths. One night in particular stood out in Pem's memory—a brisk, moonless night in January when she and Phil were taking the late ferry back to their cottage on Derby Street.

They'd walked out on the deck, and Phil pointed out some of the constellations and planets he had learned from his father. Then, out of the blue, Phil said "Some day, I'm going to build a space ship and go out there—and I hope that you will want to go with me."

A chill ran down Pem's spine. She'd never been off the ground, never even been up in an airplane. Now, suddenly, the man she had married, a man she knew was quite capable of achieving any fanciful dream he imagined, was proposing to take her into the infinite darkness of space.

The very thought of it scared her witless. She was silent for a full minute, until Phil asked her, "Does that scare you?"

"Yes," Pem answered, "it scares me to death. But I'm not going to let you go off into space without me. I get goose bumps just thinking about it, but I suppose I'd rather die with you in space than live on Earth without you."

"That's my girl," Phil said with a smile, "That's what I wanted to hear. But you can relax, we've got a lot to do before we could take on such a project, and it may take longer than we think to make something commercial out of television."

Pem squeezed Phil close. "Trust *you* to think big."[12]

After a few months of patient study and practice, Cliff felt he had grasped the fundamentals of the glass-blower's craft, and the business of fabricating the world's first electronic television camera tube began in earnest. Unfortunately, Cliff's earliest attempts at shaping the glass just the way Phil wanted it might have been comical if they hadn't been so maddening. Cliff was using a blowtorch to heat and contour the glass, but found it extremely difficult to cool the glass evenly. Every time he put a new tube on the vacuum pump, the glass would crack.

After a few frustrating attempts, Phil was fortunate to run into Bill Cummings, who ran the glass lab at the university at Berkeley. Phil described the problems Cliff was having, and once again, providence seemed to intervene in the form of an offer. Cummings said, "Bring the parts over to my lab and I'll make you a tube." That night, Phil and Cliff brought him the guts of the tube, and the next day Cummings brought the finished product to the lab and helped Phil and Cliff get it on the vacuum pump. With Mr. Cumming's help, they finally had themselves an Image Dissector.

That wasn't the only difficulty the boys encountered at this early stage. Creating the photoelectric surface of the Image Dissector proved equally daunting. A very rare substance called cesium was chosen to form the surfaces that would perform the miracle of converting light into electricity. Typical of the resource-fulness their work required, the boys struck upon a novel way to acquire enough of the precious substance. Small pellets of cesium were often used in radio tubes to absorb any residual gasses that remained after all the air had been pumped out. So the boys pur-chased whole cases of radio tubes, took them up to the lab, and smashed them with hammers to retrieve the cesium pellets.

In this painstaking manner, with invention often preceding in-vention, Phil and Cliff spent a year laying the foundation for televi-sion. The Image Dissector went through numerous incarnations before circuit tests began to show that there was indeed some kind of electrical current coming off the cesium surfaces. After that, every new tube they made was going to be "the one" that would finally give them some kind of picture. And every time they dis-covered that they still had a lot more to do before they could reach that much-anticipated breakthrough.

As soon as Phil and Cliff began to detect the faint traces of a signal from the earliest Image Dissectors, they encountered another of the "great difficulties" that Mott Smith—the engineer that Leo-nard Lyon had consulted in Los Angeles—had predicted they would encounter on the path to perfecting Phil's invention—amplification. They found it very difficult to amplify the signal coming from the cesium surfaces without also increasing the noise inherent in the circuits. To help sort out these problems, Phil added the first of many new members to the lab gang at 202 Green Street by hiring a University of California engineering graduate named Carl Christensen.

In Christensen, Phil found the expertise he needed for resolv-ing the amplification problems. He also encountered the proto-type for a character that would reappear throughout his career—the resident skeptic. There is no doubt that Christensen was an able engineer, but he also "harbored grave doubts"[13] about the feasibility of Phil's ideas, and frequently expressed his reserva-tions. From this doubting soul, Phil learned perhaps as valuable a lesson as he would ever learn from the tubes and circuits on his workbench—that he would not always find himself in the company of colleagues who shared his vision, that not everybody he worked with would faithfully follow him on his relentless march into that unseen world beyond the scientific horizon.

Carl Christensen, lab gang member and resident skeptic

Through the summer of 1927, Phil's little lab gang began to grow. In addition to Carl Christensen, Phil hired his cousin, Arthur Crawford, to help with a variety of tasks around the lab. And, along with her chores as the resident draftsman, Pem learned how to use the spot-welder and took on some of the more delicate metalworking jobs. Progress was palpable. Soon it would be time to attempt to transmit an actual picture.

Whenever he was in town, Les Gorrell would stop by the Green Street lab and inquire jokingly, "Well boys, have you gotten the damn thing to work yet?" And Phil would try to explain what he was up to, but the details were usually beyond Les.

With less than a month left in the twelve-month window of opportunity the bankers had promised, Phil and Cliff rigged together a rudimentary apparatus and began testing it to see if the system could send an image from the camera to the receiver. The first few tests revealed very little. The receiver glowed when the current flowed through the cathode ray tube, but Phil couldn't see anything except electronic interference on the screen. Ignoring his discouragement, he analyzed the results from each test and redesigned parts of the system.

On September 7, 1927 the system was ready to be tested again. This time Phil was so confident that he invited George and Pem to the lab to see the crucial experiments.

For his test that day, Phil chose the simplest of images—a thick straight line painted onto a glass slide. While this seemed like a mundane choice for such an auspicious occasion, it was in fact exactly what Farnsworth needed: if he could tell by looking

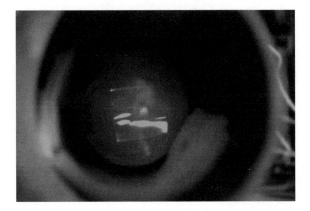

The first electronic television picture (recreated in 1977)

at the receiver whether the line was vertical or horizontal, then he could be certain that information was actually being transmitted from the bottom of one bottle to the bottom of the other.

Early that evening, Cliff Gardner dropped the glass slide between the Image Dissector and a hot, bright, carbon arc lamp. In the adjacent room Phil, Pem, and George watched the face of the receiver as it flickered and bounced for a moment. When the system settled down, all present could see the straight-line image shimmering boldly in an eerie electronic hue on the bottom of Farnsworth's magic tubes. When Cliff rotated the slide, everybody could see the image on the receiver rotate as well, clearly proving that they were witnessing the transmission and reception of visual intelligence. The Age of Information was born.

Ever-doubtful Carl Christensen, was for once genuinely impressed. "If I wasn't seeing it with my own eyes," he offered cautiously, "I still wouldn't believe it."[14]

Later that evening, Philo T. Farnsworth recorded the arrival of true video with a simple scientific statement in his laboratory journal: "The received line picture was evident this time."

In a telegram to Les Gorrell in Los Angeles, George Everson put it much more directly: "The damn thing works!"

FIVE

Something a Banker Can Understand

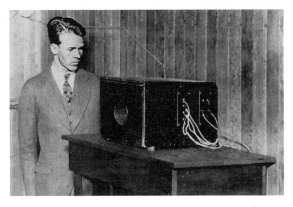

Cliff Gardner with a TV receiver, ca. 1928

"Great spirits have often encountered
violent opposition from weak minds."
—Albert Einstein

In the fall of 1927, Farnsworth and his friends became the first
humans to gaze into the shimmering eye of electronic television. Who could have guessed that the rest of the world would
not share the experience for another twenty years?

For Farnsworth himself, the moment of triumph quickly passed
as he considered the magnitude of the job that lay before him. A
crude, flickering image proved that the idea he'd nurtured since he
was a teenager would actually work as he'd imagined; it also
proved what a staggering amount of work would be required to
take this fragile invention from the laboratory to the living room.

There were occasional callers at 202 Green Street in the
weeks and months after these first successful experiments. Some
of the backers, most notably William Crocker, took a keen per-

sonal interest in the venture and dropped by from time to time to see what, if anything, had become of this curiosity they'd launched. Invariably, he'd arrive when the equipment was disassembled, strewn across the workbench and "temporarily inoperative." That was fine with Crocker, who would perch on a stool and watch the men work, asking occasional questions about what they were doing.

The other backers had only George Everson's eyewitness accounts to assure them that Farnsworth had indeed produced meaningful results. With occasional reports from George, the backers were content to let the work continue, though Daddy Fagan would often ask George pointedly, "Seen any dollar signs in that boy's gadget yet?"

Work continued, funded for the most part out-of-pocket by that syndicate of wealthy capitalists frequently referred to as "The Crocker Group" or "the bankers," although the affiliation with the Crocker Bank was indirect. These financiers had personal and business ties to the bank, and Roy Bishop was their point man on this particular project. But there was really no "Crocker" money involved; it was private money. Willie Crocker, Roy Bishop, Daddy Fagan, Jess McCargar, and the others were, in modern parlance, "angel investors." Each was sufficiently wealthy that together they comfortably accommodated the escalating cost of their speculation in television. But even the richest of capitalists has a limit.

By the end of 1927, the concept of "television" was becoming something of a hot topic in the press and in financial circles, as some of the competing mechanical systems made headlines. John Logie Baird, an eccentric Scotsman working alone out of London, had actually succeeded in transmitting a reasonable facsimile of a moving picture across the Atlantic in 1926, and was selling "Televisor" kits in England. Stateside, C. Francis Jenkins, who had been a pioneer in the development of motion picture cameras and projectors, had introduced his "Radiovisor" system. Jenkins had no problem selling stock in the venture, although he was not yet selling any actual receiver kits.

Not to be outdone by maverick inventor types, the well-funded industrial research labs were hot on the trail of the new medium as well. At General Electric in Schenectady, New York, the respected Dr. Ernst Alexanderson developed a sophisticated mechanical television system and received some press when he broadcast the first play staged for television. A one-act melodrama, "The Queen's Messenger" lent itself perfectly to the task, with two immovable mechanical scanners focused on the only two characters in the script, both of whom sat fixed in their chairs.

At the Bell Labs division of AT&T, in April 1927, Dr. Herbert Ives transmitted fuzzy but recognizable pictures of then-Secretary of Commerce Herbert Hoover from Washington, D.C. to New York, employing what has to be the prizewinner for the most bizarre mechanical system of them all. Ives' system employed an electric motor rigged to an ungainly harness of wires that sent an electrical impulse to each of twenty-five hundred individual bulbs arranged in a grid-like receiving screen. Despite the marginal nature of the technology involved, the *New York Times* reported the event as if it was one of the great breakthroughs in human history.[15]

What all these experiments had in common was their reliance on some variation of the old Nipkow disk for their scanning apparatus. True, these systems produced the best results that mechanical systems had produced to date. Most important, they produced the best results such systems would ever achieve— which is to say, the results were pretty dismal.

Out in San Francisco, Phil Farnsworth kept close track of these reports, even subscribing at one point to a whole series of Bell Labs journals. So he knew exactly what the competition, if you could call it that, was up to. The term applies only loosely, however, because it was clear that none of these operations had anything even close to what Farnsworth had. If this was a horse race, as Roy Bishop had implied back in the summer of 1926, then Farnsworth had already lapped a field that had yet to even leave the gate.

Herbert Ives and the AT&T system, April 1928

The importance of this kind of lead cannot be overstated. In fact, the measure of Farnsworth's lead over other contenders in the field was unprecedented in the annals of invention. His lead was a matter of magnitude, not degree. The others were not even on the same track, let alone following or threatening to overtake him. Farnsworth would in fact have the field of electronic television entirely to himself for nearly three years. Such an accomplishment stands in stark contrast to the experience of, for example, Alexander Graham Bell, whose patent application for a telephone device preceded a similar application by slightly more than three *hours*.[16]

What this lead meant to Farnsworth and his team of dedicated and loyal workers was that they could scoop up all the new inventions they stumbled across as they built upon their initial success. Before long they'd have a veritable sack stuffed full of valuable patents. Phil could certainly see the dollar signs in his tubes, even if Everson and Fagan could not.

Despite the doomed future of the technology behind all the mechanical television experiments, the publicity they generated enhanced the activities at 202 Green Street in one critically important way: It helped the backers sell stock to raise more funds.

The Crocker group grew steadily as new investors were enlisted to raise additional capital to meet the growing monthly expenses. When the initial $25,000 stake was exhausted, the investors invoked a provision requiring all the partners to contribute proportionately in accordance with the size of their holdings. Most of the backers were experienced investors with enough cash on hand to meet their obligation without diluting their equity, but Farnsworth was both the largest single stockholder and the least able to draw on personal cash reserves. Consequently, much of the stock that was sold to new investors was Phil's.

Along with the roster of investors, the workforce at 202 Green Street grew steadily. All of the men Phil hired were caught up in the pioneering spirit of their work and shared a special camaraderie. As one worker put it, "You never had the sense that you were working *for* Phil. It always felt like you were working *with* him."

Besides Phil and Cliff, Robert E. "Tobe" Rutherford worked on circuitry, Harry Lyman researched radio engineering and broadcasting issues, Harry Lubcke worked with Phil on mathematical problems, and Thomas Lynch was the resident machinist. And never far from the heart of things was Pem, Phil's "personal secretary, draftsman, friend, confidant, and lover."[17]

Throughout the fall of 1927 and into the winter of 1928, Phil and Cliff and the lab gang disassembled, modified, and reassembled the system dozens of times. Those first successful experiments in the fall displayed approximately fifty lines per frame, at a rate of ten frames per second. True, these results were comparable to those generated by the mechanical systems that transfixed the popular press during this period. The difference, of course, is that a blurry fifty lines per frame was the level of clarity at which the mechanical systems reached their zenith, while this was the resolution at which Farnsworth was starting.

With his basic principles now proven, Farnsworth concentrated his attention on increasing the number of scan lines, setting his sights on a minimum of 400 lines per frame, and at least

24 frames per second—the same number of frames the movies use to fool the eye into seeing a seamless motion picture.

In the weeks immediately following the September 7 experiments, Phil and Cliff continued using geometric shapes such as crosses and triangles for their experimental transmissions. One night while looking at the receiver, Phil saw a moving line that looked like...uh oh...smoke. The moment he realized what he was seeing, he thought the lab was on fire. In the next instant, Cliff's hand waved in front of the camera—holding a cigarette. Phil sighed gratefully and peered closer at the swirling patterns. He could clearly see the delicate spirals of vapor; the whole effect was conveyed with startling definition.

The first time George Everson witnessed images of smoke for himself, he declared it was a sign of real progress, though still he had no "dollar signs" to report to Fagan.

Fagan could have joked that the only progress at 202 Green Street was dollars going up in smoke, but in fact the period immediately following the first transmissions proved incredibly prolific for Farnsworth and the lab gang. Their numerous advances demonstrated just how fertile this new ground was that Phil Farnsworth was cultivating.

Although many aspects of the earliest images needed refinement, the most persistent challenge Farnsworth faced was the matter of amplification. The problem was doubly vexing. First, he needed to increase the strength of the signal coming from the photoelectric surfaces of the Image Dissector. Then, in order to increase the number of lines per frame and the number of frames per second, he needed to amplify that increased signal strength across the ever-widening range of frequencies, or bandwidth, that video requires.

Ultimately, what they needed was an amplifier that could deliver a reasonable signal-to-noise ratio across a very large bandwidth; they needed a "high gain, high bandwidth" amplifier tube. The problem was, there was no such animal. Prior to this work, there had been no need for one. They had to make their own.

Most electronic tubes of the day consisted of three elements, and were called triodes. The Farnsworth gang started fabricating their own pentodes—the first vacuum tubes with five elements. The extra elements seemed to do the trick, boosting the signal strength, while improving both the signal-to-noise ratio and the range of frequencies that could be amplified. All they had to do was create their own sockets for mounting the tubes into the transmitter chassis. Such was a typical day in the life at 202 Green Street, where not only the art of television, but also the whole science of electronics was lifted to new levels of proficiency.

Further evidence of the rich vein the lab gang was mining came out of their amplification work. In one truly groundbreaking set of experiments, the team employed a specially modified tube they called the Image Analyzer, to actually watch its own photoelectric layer being applied, practically molecule by molecule. The process used the new amplifier circuits to magnify the image electronically to such an impressive degree that the workers could observe surface details not visible to the naked eye or with conventional instrumentation.

One visitor to the lab observed this technique and said, "I don't know too much about this television work or what that's likely to lead to, but *this* is really important."[18]

What this visitor was commenting on was a procedure that would eventually evolve into the modern electron microscope, one of the most valuable scientific tools ever devised. The rudimentary foundations of the electron microscope were first discovered by Farnsworth and his colleagues, although it was left to others to further the art and go through the door that was first opened at 202 Green Street. Farnsworth's operation simply lacked the resources, the money, and the man-years needed to pursue this particular line of research.[19]

Elsewhere in the lab, Cliff Gardner was making his own great strides in the fabrication of the Image Dissector. For months, the experts had been telling Cliff and Phil it was impossible to seal a flat, optical lens on the end of a glass cylinder, but that is precisely what the transmission of television images required. When he started, Cliff was making tubes with a blowtorch in one hand and an air tube between his teeth. With such a rig, it was difficult to construct anything fancy out of glass.

Eventually, Cliff came up with the idea of mounting a turntable inside a small furnace and placing the tube and optical glass within the furnace to heat them evenly. "I got so that by turning the work slowly," Cliff recalled years later, "That I could seal the tube in only one time around."[20]

With flat glass sealed on the ends of his camera tubes, Farnsworth eliminated another source of distortion in his television pictures. Yet, looking more closely at those pictures, he observed ghosting, smudging, and a variety of other electronic artifacts that marred the image on his receiver tubes. Working with Harry Lubcke and Tobe Rutherford, he began to plant the seeds for some innovative scanning and synchronization techniques that would eventually become fundamental components of the Farnsworth patent portfolio.

He also began studying a phenomenon called "secondary electron emissions," which caused certain kinds of metal surfaces to release a cloud of new electrons when bombarded by primary electrons. Observing this phenomenon, Phil suspected, would prove even more useful in the development of future amplifiers.

In other words, with their first "strike" in September of 1927, the lab gang had opened the mother lode of ideas that would define the art of electronics for years—if not decades—to come.

As the months passed, George continued making his periodic reports to the bankers, and Fagan continued to tease him about whether he'd "seen any dollar signs yet?" Finally, in the summer of 1928, Roy Bishop pulled George into his office and handed him some figures. They showed that more than twice the original limit—nearly $60,000—had been spent to meet lab expenses. Confronted with these figures, George agreed with Bishop that it was time for Farnsworth to show his invention to the people who were paying for it.

Phil was just as surprised as George was at Bishop's figures. Still, he was hesitant to make a formal presentation of his work;

the system was still very fragile. Its reliability was tenuous, at best. Besides, he felt that he was on the verge of producing a really fine picture, one that would be clearer and much more stable. He pleaded with George for just a few more weeks, but George reminded Phil of the serious tone in Bishop's voice, and a date was set for a demonstration.

The Crocker group reassembled at 202 Green Street in August 1928. They were together for the first time since that unlikely day nearly two years earlier when they first met a twenty-year-old boy who told them he could invent television. They understood very little of what Philo Farnsworth had told them that day, but for some inexplicable reason it smelled like a winner, so they had gone for it. Now, as they were ushered into the quiet, darkened lab, they had no idea what to expect.

Phil had Cliff create a very special slide to use in this first demonstration for the backers.

"This is something a banker can understand," Phil said to George, evoking Fagan's frequent query as he switched on the system. When the little round screen hummed and flickered to life, Fagan and his friends gazed in hushed amazement as an apparition of a dollar sign ($) materialized out of the darkness.

Roy Bishop was the first to speak after the demonstration, which had also included a variety of geometric shapes and the swirling kinetic patterns of cigarette smoke. Bishop warmly congratulated Phil for delivering "his end of the bargain" and then set a more sober tone for the weeks, months, and years ahead.

"It will take a pile of money as high as Telegraph Hill to carry this thing on to a successful conclusion," Bishop declared. Phil listened anxiously as Bishop added, "I think we should take immediate steps to sell this invention to one of the large electrical companies that can afford to provide more adequate capital and facilities."[21]

Bishop's proposal came as no surprise to Phil. He fully expected all along that once he had proven that his idea would work, the backers would try to sell it quickly, hoping for a handsome cash return on their original investment. Similar scenarios

were common to the stories of many other inventors, but Phil
Farnsworth was determined not to share their fate. He had done
all he could to forestall this moment of reckoning by postponing
the demonstration as long as possible. Now he tried to forestall
the moment a bit longer as he contemplated his response to
Bishop.

When he finally spoke, Farnsworth's carefully chosen words
concealed his true anxiety. Calling once again on the charisma
that had originally gotten him where he was, Phil addressed his
sponsors, outlining what had been brimming through his own
mind, his own scenario for the future.

"Mr. Bishop," Farnsworth began, "I don't think you fully appreciate the enviable position that you and your partners are in.
You...we...have broken ground in fertile new territory. We already have critical patents pending. If we continue to refine our
system, many more patent-worthy improvements will follow.
Eventually, our company... *your* company...will compile a portfolio of patents that will control an entire industry. Anyone who
wants a seat at the table will have to pay us for the privilege."

There was no question, judging from the results to date, that
many problems were yet to be solved. By continuing at their present pace, the clever men of Farnsworth's lab gang would find the
problems first, solve them first, and file patents on those solutions.

"What you see here is just the beginning," Farnsworth continued. "Every problem we confront, every solution we come up
with will extend our lead. If we can continue at our present pace,
it will not be long before we've wrapped a hammerlock of patents around this invention. When that has happened, then we
will earn infinitely more in royalties from licensing those patents
than we will ever get if you cash out now."

Phil couldn't argue Bishop's premise—the real work of refinement and engineering was just beginning and the work
would be costly. But this was not a tangle of tubes and wires ly-
ing on a workbench—this was true television, the ultimate fulfillment of mankind's most enduring dream—to see beyond the
horizon. Compared to the mechanical systems of the day, this

was the infinitely superior mousetrap, and it seemed like only a matter of time and perseverance before the entire world would beat not a path but a veritable highway to their door. And that highway, Farnsworth was certain, would be paved with gold in the form of patent royalties.

With some verbal assistance from George, the matter of selling the venture was tabled—temporarily—and the Crocker group agreed to continue finding money to support Phil's work. But there was little time for comfort. Farnsworth realized that there was a genuine disparity between the nature of his own intentions and what he could expect in the future from his present backers. From this nest of circuits and tubes and wires, Phil envisioned a golden goose that would support his research for years to come. His investors, it seemed, were more interested in today's goose eggs.

After Bishop and Fagan and their cronies left the lab that night, Phil and Cliff sat down together for a moment to think about what had happened. As they talked, Phil contemplated the implications of what Roy Bishop had proposed; the prospect of selling out left him numb and cold. He could not fathom the prospect of working for anybody but himself. He wanted—he needed—to be his own boss, to set his own pace, and to exercise the freedom in the years ahead to follow his own imagination.

Thus, in the aftermath of the first demonstration for his backers, Farnsworth entered a new phase in his life and career. A boy's dream to make an idea work had become the classic struggle of a gifted young inventor to continue his pioneering work while also maintaining the very independence that nourished his original thinking.

However noteworthy the success he'd achieved may have been in 1928, Farnsworth was inventing much more than television. He was creating more than camera tubes, amplifiers, or waveforms. He was chasing a much more noble goal than simply refining pictures as they danced through the ether. Philo Farnsworth was engaged in the time-honored American tradition of inventing himself. The future that he envisioned for himself went far beyond the future that he foresaw for any of his inventions, to a place where his divinely inspired imagination could flourish, where his inventions would reveal new insights into the inner workings of the physical universe, which in turn would yield more inventions.

Sadly, that was something a banker could not understand.

Six

Out of the Ashes

Farnsworth with an early Image Dissector

"Success is how high you bounce when
you hit bottom."
— General George S. Patton

R eaders of the *San Francisco Chronicle* added a new word to their vocabularies when they read the headline spread across the front page of the second section on the morning of Monday, September 3, 1928:

SF Man's Invention To Revolutionize Television

Hoping to capitalize—literally—on some of the publicity that mechanical television had generated over the past year, Farnsworth had, at his backers' urging, opened his lab to journalists two days earlier. The Saturday demonstration hadn't attracted many individual reporters, but the story was picked up by the Associated Press and generated similar headlines around the country.

Under a subhead that read "New Plan Bans Rotating Disc in Black Light," the accompanying text described how the "young genius" produced a "queer-looking line image in a bluish light which smudges and blurs frequently, but the basic principle is achieved and perfection is now a matter of engineering...."

A front-page photo of Philo T. Farnsworth, sporting a mustache he hoped would make him look older than his twenty-two years, and posing as he would a hundred times more with his magic jars in hand, appeared alongside the article. The Image Dissector was described as being the size of "an ordinary quart jar that a housewife uses to preserve fruit." The article went on to say "the system is simple in the extreme, and one of the major obstacles to the perfection of television is thereby removed."[22]

Phil and Pem, along with Cliff and his own bride, Lola, picked up a copy of the *Chronicle* from a newsstand on Market Street and momentarily basked in the glow of their sudden fame. Then Phil chimed in with a sobering thought.

"This leaves us wide open to our competition," Phil reflected, his casual tone masking his genuine apprehensions. "We're still years ahead of the pack, but our inadequate financing means that we will be working under a severe handicap." Then, trying to paint a more optimistic picture, he expressed precisely what made the Farnsworth operation unique: "We have something the big companies don't have. Our small size and method of operating allows us to maneuver like a speedboat alongside their juggernauts. But even speedboats eventually run out of gas. We have our work cut out for us, that's for sure."[23]

The sudden flurry of publicity surprised no one, least of all the backers, some of whom had begun courting the press in anticipation of refinancing the enterprise. However, despite the favorable press, Farnsworth's financial underpinnings were more shaky than ever. The bankers and industrialists who funded his initial experiments had by this time been out on a limb for more than two years, and were beginning to chafe at the prospect of another year of expenses.

Unknown to the inventor at the time, considerable maneuvering was going on behind the scenes. A few months earlier, in the spring of 1928, Jess McCargar's association with the Crocker Bank had been unceremoniously severed. The precise reason for the dismissal was never explained, except for one of the other investors telling Farnsworth in a private aside, "We didn't let him go for nothing."

Whatever the reason, McCargar's departure from the bank caused a rift among the original backers. Farnsworth's fate now hung between two factions, one led by McCargar and George Everson, the other by Roy Bishop and the Crocker group.

Young Genius and Part of His New Black Light Machine

Philo T. Farnsworth holding the sending and receiving tubes of his new television set.

Farnsworth shows off his "magic jars"

Phil knew that changes were imminent when the operating funds for the lab were abruptly shut off. Under these conditions, he was faced with the unpleasant task of dismissing some of his men, while the few that remained agreed to accept a minimum living wage. Farnsworth rose to the task reluctantly, for he was being asked to lay off the only people in the world who really understood what he was doing and the way he was doing it. He had trained most of these men personally, and he felt that, as an investment, they were worth much more than the wages that they were paid. The lab gang was an invaluable resource, the sorcerer's apprentices that made all the magic happen.

Phil assured everybody that he would rehire them just as soon as the financing was straightened out. In the meantime, the

Green Street lab shifted its priorities from perfecting television to taking on freelance electronics projects on a work-for-hire basis.

As if the apprehensions about the press exposure and the uncertainties about further financing were not enough to rattle the foundations of Phil's fledgling enterprise, later that fall a real disaster struck.

Early one Sunday morning in November, Pem and Phil had just squared off against Cliff and Lola for their weekly tennis match when a policeman showed up courtside.

"Are you Farnsworth?" the policeman asked Phil, interrupting their game. Once the man with the tennis racket was positively identified, he was informed, "You might want to get down to your laboratory right away... the place is on fire."

As Pem recounted, "We rushed over... it was pretty well over when we got there, but the firemen were all spooked, because every time they tried to pour water onto the flames, something would explode. There must have been a vacuum pump working. The tube would have burst, and there were chemicals like potassium and sodium, things of that sort that would just blow up when they came in contact with water."

The fire chief informed Phil in no uncertain terms, "One more explosion and we're not going back in there!" But there was little anybody could do at that point. The damage had been done. The lab "was a shambles... the fire and the water had pretty well wrecked the contents of the lab," and Pem feared she was witnessing "the final end of our beloved television project."[24]

The devastation caused by the fire underscored the formidable hazards involved in Farnsworth's line of work. Some of the chemicals they used, such as potassium, were highly volatile, to say nothing of toxic. On one previous occasion Cliff had narrowly avoided losing his eyesight when a vial of potassium nitrate exploded in his face. Vacuum tubes were very fragile and would occasionally implode without warning, hurtling tiny shards of glass in every direction. And there was always the lingering possibility that someone would touch the wrong terminal and get a jolt from the ever-present strong currents and high voltages. These dangers were a constant presence in the day-to-day work at 202 Green Street.

The men considered themselves fortunate that none of them was ever seriously injured in the course of their daily tasks. But the fire seemed to outweigh all that good fortune. Any number of things could have caused the blaze, but the fire department investigating the scene never did find a specific cause to report, other than to offer the obvious assessment that they suspected "something electrical."

For one sleepless night, Phil and Pem worried about whether, amidst all the other financial uncertainties hovering over the lab, the insurance premiums had been paid. They were greatly relieved to learn the next morning that everything was in fact covered. That very Monday, the lab gang set about the messy task of recovering their work out of the ashes and water-logged remains.

They immediately seized on the opportunity to start again from scratch, using the insurance proceeds to apply everything they had learned in the past two years in constructing a completely new setup. Phil put his men to rebuilding the system using only their most recent circuits and tubes, and when the first tests of the rebuilt system were conducted a few weeks later, "the picture was better than ever."

While the gang was rebuilding, Phil had a chance to turn his attention to some of the ideas that had been simmering in the back of his mind, particularly the scanning and synchronization questions he had identified earlier. He knew that before he could measure any further progress, he needed an image source with more detail than he could render using simple geometric shapes painted on glass slides, or the curling traces of smoke that had conveyed the first images of motion. They tried to televise live human subjects, but the extremely hot lights still required for the Image Dissector prevented the use of live models for more than a few seconds. They had to come up with something else.

An entry in the lab journals notes "transmission of motion pictures by electrovision attained by Rutherford." In other words, Tobe Rutherford had developed what is known as a "film-chain," a device for displaying motion pictures on television. One observer commented wryly that the journal entry just meant "they bought an old projector and got it working."

The first film chain, ca. 1929

But there was really much more to it than that. Tobe had to reconcile the different frame rates and eliminate the black flicker caused by the shutter of the motion picture projector.

And motion picture projectors were not cheap. The purchase put the operation well over budget for the month, further testing the patience of the backers. But once Phil and his assistants got the film-chain working, they had the means to really see the progress in their pictures. They obtained some film loops, including a clip of a new cartoon character called Mickey Mouse in an animated short called "Steamboat Willie," and a loop of the same Dempsey–Tunney "long count" title fight that Phil and Pem had heard on the radio the night they arrived in San Francisco.

By far, their favorite film loop was a clip of Mary Pickford, "America's Sweetheart," combing her hair in a scene from *The Taming of the Shrew.* George Everson quipped, "Mary Pickford combed her hair at least a million times for the benefit of science and the development of television." As the details in Miss Pickford's features and the lovely glowing strands of her hair were gradually refined, the team could see much more clearly the obstacles that remained on the path toward producing a commercially viable television picture.[25]

To understand the problems Farnsworth now faced, it might be useful to first explain in more detail precisely how electronic television works.

The process is easiest to visualize in a television receiving tube, where an electron beam acts like a needle-thin paintbrush. The intensity of the electron beam fluctuates according to the brightness of different elements of the picture, causing the phosphors on the inner surface of the tube to glow in relation to the intensity of the beam.

The tube is surrounded by a harness of electromagnetic coils. The first set of coils, on either side of the picture tube, deflects the electron beam horizontally, bouncing it back and forth across the screen so that the phosphors are illuminated one line at a time. At the end of each newly scanned line, the beam retraces its path across the tube to repeat the process again with the next line, until the beam has painted hundreds of lines per frame.

Simultaneously, another set of electromagnets, situated above and below the tube, deflects the beam vertically, drawing the electron stream downward across the image, so that each horizontal line appears below the line above it. When the vertical scan has reached the bottom of the frame, the beam retraces its path to the top of the screen, where the process begins all over again with the first line of the next frame.

The entire process is repeated dozens of times per second. Just as with motion pictures, each scanned image persists within the eye just long enough to create the illusion of smooth motion.

The scanning process is reversed in the video camera tubes. When light is focused on the photoelectric surface of the camera, it emits the electrons that form the beam, which are then focused and manipulated by the magnetic fields around the tube in much the same manner as in the receiver.

This is the essence of the process that Philo Farnsworth first demonstrated at 202 Green Street in September 1927. After rebuilding the lab, and with the images from the film-chain offering more detailed subject matter, a number of defects in the picture became much more apparent. As George Everson described it, one of these defects was "blurriness...due to a double image" creating "the appearance of a shadowed reflection." The other prominent defect was a vertical "black splotch" through the center of the picture "as if someone had taken a dirty finger and smudged it from top to bottom."[26]

Both problems, Farnsworth suspected, were caused when the electron beam retraced its path across the tube. They were possibly caused by the actual "shape" of the electrical impulses that produced the magnetic deflection, called a "waveform."

Phil demonstrated the idea for George one day. "If you look closely, you'll see that the scan lines have a sort of circular path," Phil explained. "The ghost image is in the return lines. We can get rid of it by straightening out the scanning pattern, and by shortening the time of the return scan. That should take care of our double images."[27]

In slightly more scientific terms, what Phil was showing George is called a "sine wave," the most common electrical waveform. A sine wave consists of smoothly shaped crests and valleys suggestive of waves or ripples on a body of water. The up-slope of the wave is a perfect reflection of the down-slope. Thus, the return scan produced a duplicate of the primary scan, causing the ghostly reflection.

What the situation called for was a way to scan the image so that the primary scan would do its job, with the return scan being very fast, almost instantaneous, so that the electron beam would begin scanning the next line before the retrace could leave an impression on the screen. In other words, from a wave shaped like ripples on the surface of water, Farnsworth needed to come up with a wave shaped like the teeth of a saw, with a straight, gradual up-slope, and a severe, almost vertical down-slope. That's precisely what Farnsworth came up with, and a "sawtooth" wave is just what he decided to call it.

The new waveform proved to be a tremendous advance in the refinement of electronic television. As George reported,

"Gradually the smear down the center of the picture disappeared, the ghost image faded out, the picture field cleared, and the image became sharper."[28]

Once he had perfected the sawtooth wave, Farnsworth devised a companion scanning technique called "blacker than black," which caused the electron beam to virtually extinguish itself during the return scan, further eliminating any trace of ghosting. As soon as these new techniques were demonstrated, it was time to file more patents.

Here we see the real dreams of Philo T. Farnsworth unfolding. This is just one example of the grand scheme that Farnsworth had tried to map out for Roy Bishop and the backers: A breakthrough invention like electronic television leads to new insights, which lead to new inventions. The race to invent television, from the inventor's point of view, had become a matter of refinement and engineering. The real challenge at that stage was to create the enterprise that could produce these refinements, while simultaneously nourishing the sort of unorthodox thinking that had spawned the original idea in the first place.

Invention, by its very nature, is an "outsider's" craft. The challenge every true inventor faces is to create an environment in which he can thrive outside of the existing technological and economic paradigms, because his inventions, by their very nature, ultimately threaten those paradigms and the corporate institutions built around them. The inventor's challenge is to flourish on the periphery, because that is where new truth is discovered, refined, and patented. To Farnsworth, his was a dream not of products but of a process—a dream of unfettered exploration and discovery.

By opening the door to a whole new field of investigation, each new refinement extended Farnsworth's lead over the potential competition. It also revealed more secrets that would ultimately provide still more refinements and patentable inventions. The momentum in this cycle of discovery and invention was truly palpable at 202 Green Street, as the lab gang did everything in their power to rush television to the point where it could be introduced commercially. At that point, the royalties on their patents would turn the laboratory into a self-sustaining research operation. For Farnsworth, television was the means to that end.

The principal architect in this grand design was Farnsworth's patent attorney, Donald Lippincott, who had been an unofficial, nonresident member of the lab gang for almost as long as the operation had been in San Francisco. When Farnsworth

Donald Lippincott

moved his operation from Los Angeles, the original attorneys, Lyon and Lyon, referred future filings to Charles Evans, a reciprocating firm in San Francisco, where Lippincott was an associate. In addition to his credentials as a lawyer, Lippincott was a trained and perceptive engineer, and he held Farnsworth and his accomplishments in great esteem. For his part, Phil regarded Don as "urbane without being Eastern."[29] With Farnsworth's encouragement, Lippincott started his own law firm, and the two worked closely together to build an impenetrable wall of patents around Farnsworth's inventions.

About this time, in the early months of 1929, Pem surprised Phil with the news that they were expecting their first child. With that news, and despite the difficulties and machinations going on behind the scenes, Phil cashed in some of his stock, enough to buy a new Chrysler sedan and some furniture for a house in San Francisco's Marina district. Over the next few months, a number of members of the Farnsworth and Gardner families came to live with them, most notably the grandmother-to-be, Serena Farnsworth. With the addition of a rented grand piano, life for the extended Farnsworth family seemed all that Phil had promised it would be.

While work on the picture in the laboratory was improving steadily, and as life for the Farnsworths was settling down and taking a prosperous turn on the domestic front, things were also

coming to a head in the backstage dealings between the two factions that had formed among the principal backers.

Roy Bishop and Jess McCargar mixed together about as well
as oil and water. Where Bishop was a disciplined engineer,
McCargar was more the hustling securities peddler type, "a kind
of fast character."[30] In classic Western terms, the Farnsworth venture was not big enough for the both of them. Phil might have
been better off in the long run if he could have maintained a
working relationship with Bishop, who had a firmer grasp of the
process and might ultimately have seen the wisdom in the inventor's long-range vision. There was also a strong personal bond
between the two: Bishop saw something of his recently deceased
son in the young inventor, and Farnsworth saw something of his
father in Bishop.

The pivotal player in the drama unfolding behind the scenes
was George Everson. It had been George who had first discovered young Farnsworth working for his Community Chest campaign in Salt Lake City; and it was George who had come looking
for Jess McCargar when Daddy Fagan suggested he bring his
young protégé north to meet Roy Bishop. So in the end, it was
the relationship between George and Jess McCargar that prevailed.

The behind-the-scenes maneuvering ended on March 27,
1929, with the formation of a new company, Television Laboratories, Incorporated. Roy Bishop, Willie Crocker, and the other
members of the "Crocker Group" sold their interest to the new
corporation, accepting 10,000 shares of stock in return but step-
ping aside from any further involvement in the day-to-day operations. Jess McCargar stepped in as President and Chief Executive
of the new enterprise, Farnsworth was named Vice President and
Director of Research, and George Everson assumed the duties of
the Secretary/Treasurer.

To Jess and George fell the task of raising funds for the continued operations of Television Laboratories, which they proposed to
do by capitalizing another 10,000 shares of stock. The task suited
McCargar perfectly. This was, after all, the "Era of Beautiful Non-

sense" on Wall Street, and Jess McCargar was a creature of the times who had amassed a sizeable fortune as a promoter of speculations and fancy ventures.

To a man with McCargar's instincts, television was a highly promotable affair. The simple mention of the word in his circles evoked tremendous curiosity. Prospective investors always asked to see it for themselves, and once they came face to face with the electronic marvel, they were invariably impressed. Television sold itself right from the beginning, so McCargar had no trouble finding a suitable market for his stock—at least, he had no trouble in the first few months of 1929.

Farnsworth accepted the new circumstances with his usual guarded enthusiasm. He was immeasurably grateful for the opportunity to resume his work with the promise of adequate funding, and he was certain that the likelihood of an outright sellout to "one of the large electrical companies" had been averted, at least for the foreseeable future. Still, the situation was far from perfect. An unpleasant aroma lingered around all these financial shenanigans, and Phil worried that McCargar's abrasive nature might disrupt his carefully managed operations. But Phil concealed his ill ease from his friends and colleagues, again expressing to Pem only his hollow confidence that everything would work out all right.

Whatever Phil's misgivings, the great bull market of the late 1920s roared on, providing Jess and George with a willing pool of investors eager to take a stake in Television Laboratories, Incorporated. At Les Gorrell's urging, even Phil made a couple of successful stock trades, but his instincts told him that the market could not rise forever, and he ceased his speculations, grateful that he'd gotten out with some profit. Others would not be so lucky.

In September, Pem gave birth to the couple's first child. Named for his father and great-grandfather, the baby boy was christened Philo T. Farnsworth III. Six months later, Pem was pregnant again.

The Era of Beautiful Nonsense came to an end on October 29, 1929, when, as one newspaper put it, "Wall Street [Laid] an Egg." Only time would tell if the impending financial maelstrom would include the dreams of Philo Farnsworth among its casualties.

SEVEN

A Beautiful Instrument

Farnsworth tunes his first prototype of a home TV receiver

"The creative inventor takes ideas out of their original
contexts and uses them in new contexts. He turns
bread-mold into penicillin, coal into electricity—or,
I suppose, lead into gold—because he isn't con-
strained to keep each thought in its own container."
—John Lienhard

In a book titled *4,000 Years of Television,* author Richard W.
Hubbell writes, "...in the 1920s, the word television became
dynamic. It took on the attributes of a lost Atlantis, a fountain of
youth, a modern Midas touch... it was *the* thing to get in on the
ground floor of...."

By 1930 Philo Farnsworth knew exactly what Hubbell was
talking about. He saw the evidence firsthand, as a steady stream
of visitors needed to look no farther than the image on Farnsworth's magic tubes to realize that 202 Green Street was the
home address of "the ground floor." Despite chaos in the financial markets, interest in television continued unabated, and Phil

spent almost as much time demonstrating television for potential investors as he did actually working in the laboratory.

In early March, Phil learned that a valuable if unlikely potential ally wanted to visit the Television Laboratories.

The United Artists Group was one of Hollywood's premier movie operations, and they wanted to see with their own eyes this new medium they had been hearing about, to assess for themselves its possible impact on their industry. Perhaps they would also make a substantial investment in the newly incorporated laboratories. As Pem put it, "If they were going to be put out of business, they wanted to be part of the act that was coming in."[31]

United Artists consisted primarily of producer Joseph Schenk and two of Hollywood's most popular stars, Douglas Fairbanks and his wife, Mary Pickford. Yes, the very same Mary Pickford who had "combed her hair a million times" for the advancement of television was about to pay a visit to 202 Green Street to see the miracle for herself. The prospect created quite a stir, as word leaked out and appeared in the San Francisco papers that Fairbanks, Pickford, and company were in town to see Farnsworth's invention.

By the time the United Artists Group called, thanks to the rebuilding after the fire and the addition of the sawtooth wave, Phil and his gang had "a beautiful picture."[32] But just before the visitors were scheduled to arrive at the lab, Phil wanted to try one more adjustment, and when he was ready to start the system again something went terribly wrong. They not only couldn't produce a beautiful picture, but suddenly they couldn't produce *any* picture at all.

The lab gang worked feverishly through the day and into the night trying to figure out what had gone wrong. The following morning most of them were still there, still with no solution. Tobe Rutherford, Harry Lubcke, Cliff, and Phil all went to Pem and Phil's place for breakfast, a shower, and a shave, and then back to the lab.

When the Fairbanks and Pickford entourage showed up at the lab that afternoon, there was still no picture. Embarrassed, Phil, Tobe, and Harry continued poring over the system, and Cliff was asked to entertain the visitors by demonstrating some of his fancy glass-blowing techniques. Cliff thought he was doing an adequate job of dazzling his guests until Douglas Fairbanks mentioned that they'd just finished a movie for which they'd brought in glass-blowers to create some special effects, so Cliff wasn't showing them anything they hadn't seen already.

The visitors were gracious and patient, but when it was time for them to leave, they still hadn't seen a television picture. Shortly after they left, Tobe Rutherford discovered that the problem was just one little wire that had come disconnected somewhere in the circuits. Phil was disgusted by the simplicity of the thing, and apprehensive about the possible recriminations from Jess and George when they found out what had happened.

When she returned to Los Angeles, Mary Pickford wrote a nice note to Phil, thanking him for his hospitality, but it was clear by then that a vital opportunity had been lost. An investment from a group like the United Artists might have provided significant diversification to the financial underpinnings of the newly constituted company, especially as it was trying to find its footing in a rapidly deteriorating financial climate.

Still, Phil did not lose a whole lot of sleep over the unfortunate outcome of the United Artists' visit. He knew that the key to his long-term prosperity lay not in enticing more investors, but by finding potential licensees for his patents. True, the patents filed in 1927 were still pending before the Patent Office in Washington, but Farnsworth and Donald Lippincott were applying for new patents all the time, and Phil was certain that once the first patents were issued, which they expected to happen any day, he would be able to find any number of companies willing to license all his patents for generous royalties.

In that context, Farnsworth no doubt welcomed the news, in April of 1930, that Jess McCargar had arranged for him to entertain yet another distinguished visitor to the laboratory. Dr. Vladimir K. Zworykin, an engineer with the Westinghouse Company in Pittsburgh, was somebody whose experiments in television

Farnsworth was familiar with and held in some regard. His patent searches had no doubt encountered Zworykin's own pending patents for an electronic television system.

As a scientist, Farnsworth tended to accept anyone who was articulate in the subject as a fellow traveler on the new frontier, and he no doubt looked forward to the opportunity to discuss his work with somebody with Zworykin's stature. At the time, Zworykin was working for Westinghouse, the company that had employed Nikola Tesla[33] at the height of his career, when Tesla developed his system of alternating current. Tesla, with the aid of industrialist George Westinghouse, had paved the way for the electrification of America—and made both men fabulously wealthy in the process.[34] Both men were gone, but their legacy surely appealed to Farnsworth, and the company they had built was one of a handful of companies with sufficient resources to take a license that could help Farnsworth lay the foundation for another new industry.

The only problem was, when Zworykin was ushered into the loft at 202 Green Street, he wasn't really visiting on behalf of Westinghouse. He was about to begin working for another company, an outfit called The Radio Corporation of America, and its newly designated president, David Sarnoff.

Zworykin and Sarnoff were both émigrés from Russia. Beyond that fact their backgrounds could not have been more different. But it was not their backgrounds that brought them together. It was their common desire to be linked forever to the introduction of television.

Zworykin grew up among the comfortable middle class of Czarist Russia and obtained a degree in electrical engineering from the St. Petersburg Institute of Technology in 1912. While at the Institute, he paid close attention to one of his professors, Boris Rosing, who had proposed the use of cathode ray tubes to produce television images as early as 1908. Storing Rosing's ideas away for future reference, Zworykin began graduate studies at the College of France, but returned to Russia to serve in the Russian Signal Corps during the First World War.

When the Bolsheviks rose to power in 1917, bringing down the bourgeois establishment that remained loyal to the Czar, Zworykin had the presence of mind to leave his native country for the greener pastures of the United States, where his training qualified him for work as a researcher with the Westinghouse Electric Company. He also continued his climb up the ladder of mainstream science by obtaining a Ph.D.

Vladimir K. Zworykin
with Kinescope tube

in electrical engineering from the University of Pittsburgh in 1926.

Almost as soon as Zworykin started working at Westinghouse, he started pestering his superiors about television, so anxious was he to pick up where Boris Rosing had left off. In 1923, Zworykin applied for a patent for a completely electronic television system, but nearly seven years later, when he showed up at Farnsworth's lab, that patent had still not been granted. Despite an alleged demonstration of a cathode-ray television system in 1925,[35] Zworykin's superiors at Westinghouse were unable to see much promise in the work, so it was dropped and Zworykin was admonished to "work on something more useful."[36]

There is little evidence that the tubes were ever built, let alone satisfactorily operated, but the Zworykin's 1923 application date would prove pivotal in the annals of television prehistory.

Like Zworykin, David Sarnoff's family also lived in the shadow of Czarist Russia, but came to America under very different circumstances. Instead of the Bolshevik revolution, David Sarnoff's family was forced to flee violent Cossack raids on their *shtetl*, a Jewish village near the city of Minsk in what is now the post-Soviet nation of Belarus. As a child, David studied the Talmud and prepared for life as a rabbi. He abandoned those studies when the family arrived in New York in 1900 and adopted the tenement life typical of poor, Eastern European immigrants of the period.

Sarnoff began his career in the communications industry at age nine, selling Yiddish-language newspapers to residents of his neighborhood in Manhattan's Lower East Side. He taught himself English by comparing the text of discarded English newspapers to the Yiddish editions he peddled, and by the time he was ten, had learned enough English to attend the neighborhood's public school.

When he was fifteen, Sarnoff's father died, and he was forced to drop out of school and seek full-time employment to support his family. What he lacked in formal schooling he more than made up for with driving ambition. He became the messenger boy for a telegraph company, and with his first money bought his own telegraph key. He soon became proficient in Morse code and found work as a wireless telegraph operator for the American Marconi Company.

At Marconi, young Sarnoff's star began to rise, particularly when he attached himself to Guglielmo Marconi,[37] offering to be the great inventor's personal messenger and guide whenever he came to New York. With Marconi's blessing, Sarnoff became a wireless operator, and eventually was named the youngest manager in the Marconi network on being placed in command of the station at Sea Gate, New York. When he returned to the city, he was the operator of the Marconi Wireless operation when it relocated to the top floor of the John Wanamaker Department Store in midtown Manhattan.

It was at Wanamaker's that Sarnoff's biography began to part with actual history and flirt with the realm of self-serving mythology. Sarnoff was at his post the night of April 14, 1912, when a luxury liner called *Titanic* struck an iceberg in the North Atlantic. He was no doubt one of a number of wireless operators who relayed reports of the tragedy, dit-dotting over the ocean from the doomed ship and others that steamed to its aid. According to the legend, Sarnoff remained at his post for more than three continuous days, and he would later claim that he was the only wireless operator left on the air after President Taft ordered all others to cease their transmissions. There is no doubt that Sarnoff was present during the event, but many historians—except for those who served as Sarnoff's official biographers—consider his role vastly overstated.

Hero of the *Titanic* or not, Sarnoff's fate was securely tied to that of American Marconi and its various subsidiaries and successors. During World War I, the American military establishment determined that control of wireless communication was too important to be left in foreign hands, and so the assets of British-owned American Marconi were placed temporarily in the care of the U.S. Navy. When the war ended, those assets, including the American rights to Marconi's patents, were sold to General Electric, which combined its own radio-related patents with those of AT&T to form a new company, the Radio Corporation of America.

Thus RCA started out with a comprehensive pool of patents, combining those of Marconi with those of Hertz, Tesla, DeForest, Fessenden, Alexanderson, Armstrong, and other lesser known pioneers of radio, giving RCA a virtual lock on all aspects of the art and science of radio. RCA was, in other words, a government-spawned and sanctioned monopoly.

David Sarnoff aligned himself near the top of the new company. In his capacity as commercial manager, Sarnoff worked closely with the new company's attorneys to manage its patents for maximum profit. By the mid-1920s, it was virtually impossible to manufacture or sell any kind of radio equipment without paying royalties to RCA.

The company's policy toward patents was simple. "The Radio Corporation doesn't pay patent royalties," Sarnoff allegedly boasted, "we collect them." With that strategy as his prime directive, Sarnoff oversaw a take-no-prisoners policy that steamrolled dozens of smaller competitors out of business. What few patents the company did not already control, it acquired; or it sued the owners, often putting them out of business, and acquired the patents in settlement negotiations.

Through such aggressive business practices, by 1930 RCA came to control a vast array of radio-related businesses. In addition to the manufacture of all manner of radio transmission equipment and receivers, RCA's interests included ship-to-shore communications, photo-facsimile transmission, sound equipment for motion picture production and projection, phonograph equipment and recordings, and numerous other smaller services involving sight and sound.[38] The crown jewel of RCA's empire was its nationwide network of radio stations, the National Broadcasting Company. It was said at the time that a radio station owner was a king in his community, but linking together a chain of stations made "an even more resplendent emperor of him who controlled the chain." Just such an emperor was David Sarnoff, who in January 1930 was named president of RCA. That made him free to spend the company's money to advance his own grand design.

Through his ascent, Sarnoff developed a strong appreciation for the value of patents. He also came to realize that patents are a perishable asset with a limited life expectancy.

In the United States, a patent grants its owner exclusive rights for only seventeen years, after which all inventions pass into the public domain. In the late 1920s, Sarnoff must have realized that many of the patents under RCA's control would soon reach the end of their seventeen-year terms and expire. He also reasoned that some new kind of radio device might be invented that would eventually make the existing patents obsolete. Sarnoff resolved that RCA should get a handle on that new kind of radio before anyone else. That way, RCA could manipulate the introduction of the new invention in such a way as to maximize RCA's return on the old radio patents before they expired, effectively milking the existing patents for every day of their seventeen-year term—and then extending similar domination into the new service.[39]

It comes as no surprise that the new development which seized Sarnoff's ambition was not radio at all, but radio-with-a-picture. He observed that every time there was a flurry of publicity about television, sales of radio sets dropped slightly, as consumers held on to their money in anticipation of something far better. What he saw was enough to convince him that visual broadcasting would one day dwarf its sound-only predecessor. Consequently, in order to head off the threat that a new industry would eclipse his own, Sarnoff proposed to sire the new industry himself. As president of RCA, Sarnoff set his sights, and the power of RCA, on the orderly introduction of television.

And so the paths of these two Russians, Sarnoff and Zworykin, converged. When the two met for the first time early in 1929,[40] each was determined to play an instrumental role in the discovery of that "lost Atlantis"—television. Zworykin wanted to be its inventor, and Sarnoff was determined to be its controlling patron, and so reserved for himself the title "Father of Television."

When the inventor visited his potential patron at RCA's offices in New York, probably in January of 1929, they hatched a plan: First, Zworykin convinced Sarnoff that the mechanical televisions that were generating the news of the day would never work well enough to be commercially viable. Second he told Sarnoff that he had a better idea—electronic television, with no moving parts. He told Sarnoff he'd been working on it for years and was on the threshold of success. All he needed was another two years and $100,000, and he would deliver a viable electronic television system that Sarnoff could then introduce to the marketplace.

For the interim, Sarnoff agreed to provide funding for Zworykin, with the stipulation that the latter remain in Pittsburgh working under the auspices of a cooperative agreement between RCA and Westinghouse. This arrangement suited the Westinghouse people just fine, since it relieved them of the financial burdens of Zworykin's persistent aspirations. It permitted Sarnoff to venture into this costly new research arena without taking on the costs of setting up a whole new laboratory. And it also enabled Zworykin to travel without revealing the identity of his true employer.

Once Sarnoff was named president of RCA, the plan went into high gear. Shortly after his ascension, RCA completed the acquisition of the Victor Talking Machine Corporation, the company that manufactured the Victrola line of gramophones, creating the RCA-Victor Company. In the deal, RCA became the steward of Nipper, the black-and-white terrier famous for tilting his ear toward the horn of a gramophone in order to hear "His Master's Voice." RCA also acquired the Victor manufacturing plant in Camden, New Jersey, and Sarnoff decided to convert the unused portion of those facilities into a laboratory where Zworykin could pursue his own ideas.

David Sarnoff, president of the
Radio Corporation of America

Zworykin was packing his bags, preparing to move his operations from Pittsburgh to Camden in the early spring of 1930 when Sarnoff suggested he make a detour to San Francisco, to see if an upstart young inventor there had invented anything that RCA would need in order to advance its own television research. Sarnoff included one notable detail in his instructions: Zworykin was to approach Farnsworth in his present capacity, as an engineer for Westinghouse, investigating the possibility of a patent license. Zworykin's destination after San Francisco—Camden— was not to be discussed.

Why Sarnoff would want Zworykin to be circumspect about his intentions is not hard to grasp. Were he representing Westinghouse, a visit from Zworykin could easily be construed as a preliminary step toward a possible patent license. Were he representing RCA, a company known for its unyielding pursuit of patent ownership, he likely would not have been as hospitably received.

A formal examination of new patents and the work they cover is common practice in negotiating for a patent license, and would have been entirely appropriate in the course of Zworykin and Farnsworth discussing a patent license with Westinghouse. But according to witnesses, Zworykin "prowled around" Farnsworth's lab for three full days, during which time he had ample opportunity to avail himself to most of the secrets that made 202 Green Street the only address in the world with true television.

By the time he arrived in San Francisco, Zworykin had achieved some measure of success with a television receiving tube comparable to Farnsworth's. In 1929 he had introduced a picture tube he called the "kinescope," which offered some novel improvements over Farnsworth's very similar Oscillite picture tube. However, displaying a video image—converting electricity back into light—that's the easy side of the equation.

Zworykin's progress was necessarily hampered by the lack of a viable camera tube. As he was learning, the difficult part of the process—converting values of light into values of electricity—required much more than the sort of careful engineering demonstrated in the kinescope picture tube. Zworykin was still using a scanning disk for the input end of his system, and even the very best image such systems could render were limited to forty or fifty lines per frame. In other words, Zworykin could produce no more resolution in his receiver than he could generate with his transmitter. His whole system was retarded until he could come up with a better camera tube. Farnsworth's invention was the missing ingredient that had eluded Zworykin and his contemporaries.

In developing his kinescope picture tube, and relying on a scanning disk to provide the television image, Zworykin had synthesized the body of existing knowledge to come up with a modest improvement over what was previously available. In order to achieve the sort of breakthrough that would make television commercially feasible, something entirely new—a more efficient way of producing the television signal—was needed. Finding that breakthrough required a stroke of true genius, the ability to "see what everybody else has seen and think what nobody else has thought." Now Philo T. Farnsworth—twenty years Zworykin's junior—showed him what he'd been missing.

When he arrived at the Green Street loft on April 14, Zworykin was shown a clear, sharp picture with more than 300 scan lines per frame. Tobe Rutherford fired up the system for Zworykin, and as Phil recounted to Pem later, "when Tobe turned the receiver on, a group of us were in the room and you should have seen his expression when he saw the picture. His jaw just dropped and his eyes just about jumped out of his head!"[41]

That afternoon, Phil handed Zworykin the latest version of the Image Dissector, and again Zworykin was overwhelmed by what he was seeing. Many witnesses have since recounted Zworykin's response:

"This is a beautiful instrument," Zworykin said. "I wish I had invented it myself."

One of the features of the Image Dissector that caught Zworykin's attention was the flat, optical glass sealed into the end of the tube. Marveling at the construction, Zworykin said to Phil and Cliff, "My people told me that's impossible."

"That's what they told us, too," Phil replied, "but that's what we needed, so Cliff here figured out a way to do it."

Without hesitation, Zworykin asked Cliff, "Can you show me how this was done?" And the next day, Cliff demonstrated the very glass-blowing techniques that Zworykin's people had said were "impossible."

Actually, Zworykin saw much more than he had bargained for. Cliff demonstrated the method they had developed for coating the photoelectric surfaces, including the electron magnifying technique they used to observe the formation of the molecular layers. Zworykin paid particularly close attention when shown this novel procedure. He must have realized that, in addition to electronic television, he was being shown the earliest progenitor of the electron microscope.

After Zworykin's second day at the lab, Phil invited him home for dinner, where Pem served up one of Phil's favorite menus: leg of lamb, baked potatoes, sour cream and asparagus, topped off with peach upside down cake with pecans and whipped cream.

"This is how I met Zworykin," Pem recounted. "He was a very distinguished-looking man, with a very courtly, continental manner. He had a very friendly air about him. At one point in the evening, he commented to me what a brilliant man Phil was, how he would just be so happy to work with him, and I thought to myself, yes, I imagine you would...."

By the time Zworykin left the lab after his third day, Phil was beginning to have second thoughts about the whole encounter. As he thanked Phil for his time and attention, Zworykin made no mention of Westinghouse. There was no invitation for a reciprocating visit to the Westinghouse laboratory, nor any mention of their possible interest in a license for the use of Farnsworth's patents. The parting seemed ominous, indeed.

Phil was very quiet at dinner that night. "A penny for your thoughts," Pem prodded.

"I'm afraid we're letting ourselves in for trouble, by letting our techniques out of the lab," Phil answered. "Even though Dr. Zworykin seems like a very nice person, I have no doubt that he won't rest until he's getting flat-end tubes, as well as the other things we do to get our picture."

Phil grew reflective about some of the other things they had shown Zworykin. "He was impressed by our ability to watch the molecular layering of our photoelectric surfaces. As a matter of fact I'm rather proud of that idea. It has wonderful possibilities. Just think what that would do made into a microscope. If I had the time and money I would begin working on that immediately. Unfortunately, I have to keep strictly to television to maintain my lead in the field. Why, Jess and George would have a fit if I even suggested such a thing."[42]

While Phil was lamenting his inability to pursue the possibilities that appeared in his laboratory nearly every day, Zworykin wasted no time. From San Francisco, Zworykin took a train to Los Angeles, where he dictated a 700-word telegram to his colleagues in Pittsburgh detailing the specifications for a tube he wanted them to build.[43] When he arrived in Pittsburgh a few days later, his staff had his very own Image Dissector tube waiting for him.

Zworykin picked up the ersatz Dissector tube, tucked it under his arm, and boarded another train out of Pittsburgh. He was finally ready to report for duty at Sarnoff's new RCA laboratory in Camden.

EIGHT

Nothing Here We'll Need

Sarnoff and Marconi visiting an RCA facility

"Freedom is the oxygen without which
science cannot breathe."
—David Sarnoff

Some eight years after young Philo T. Farnsworth first chalked out his idea for an electronic television system for his high school chemistry teacher, and nearly three years after he proved that his ideas would work on a laboratory workbench, the U.S. Patent Office validated his work by issuing three patents in his name.

The first patent, issued May 13, 1930, covered an "Electric Oscillator System," a relatively minor component of Farnsworth's work. The second and third patents, issued August 26, 1930, were considerably more comprehensive and significant. Patent #1,773,980 covered Farnsworth's "Television System" and patent #1,773,981 covered his "Television Receiving System." After navigating a labyrinthine course through the patent office in the three years since their initial filing, overcoming numerous interference claims in the process, these patents emerged broad and inclusive. The cornerstone of the Farnsworth patent portfolio was now firmly in place—even as other bricks were being added to the structure.

Farnsworth was jubilant with the news of these first patents, as they provided a mantle of credibility that he felt was too long in coming. Likewise, his backers, principally Jess McCargar and George Everson, were greatly encouraged. Once the actual patents were issued, they were confident that they had the means to find licensees who would pay royalties for the use of those patents, and who would join Farnsworth in his crusade to deliver commercial television to the marketplace. In the meantime, George and Jess were still selling stock to raise the funds necessary to keep the lab running, a task they found increasingly difficult in the deteriorating financial conditions of the opening months of the 1930s.

Visitors to the laboratory at 202 Green Street continued. Undaunted by the curiously inconclusive result of the Zworykin visit, Farnsworth demonstrated his system for, among others, engineers from Bell Labs. This was the research arm of AT&T, a potential ally every bit as formidable as RCA. Bell Labs had conducted some experiments with mechanical television in the late 1920s, and the engineers who visited were sufficiently intrigued by what Farnsworth demonstrated for them that they invited him to visit their facilities on the East Coast. There seemed to be a genuine possibility of some kind of license. At Jess McCargar's insistence, Phil made the trip, and George Everson accompanied him.

McCargar may have had an ulterior motive for sending Phil away from the lab. Recently, Phil had hired two new hands: Russ Varian, a Stanford graduate, was an experienced chemist who was applying his considerable skills to improving the fluorescent materials in Phil's receiving tubes; and Arch Brolly, an MIT grad, who was an expert electrical engineer working closely with Tobe Rutherford in the development of new circuits. Given the difficulty McCargar was having raising funds for the payroll, and per-

haps because these two college graduates commanded a
slightly higher salary than
other members of the team,
Jess had decided that Phil
needed to pare down the size
of his staff. While he and
George were gone, Jess decided to do it for him.

As Pem recounted, "After
Phil and George left for the
East Coast, Jess went to the
lab and fired all the men, announcing that the lab would
be closed indefinitely. Then
he wired Phil and George and told them what he had done."[44]

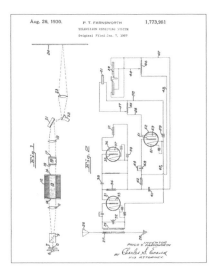

Farnsworth's patent #1,773,980

The discussions with AT&T came to an abrupt halt, as Phil
and George boarded the next train back to San Francisco. Phil
fumed all the way across the continent. To him, this summary
firing and closing of the lab was not just a draconian financial
measure. It struck straight to the heart of the pioneering scientific enterprise he was trying to build. As he had said many
times, his work was as much about "building men" as it was
about "building gadgets."

By the time they got off the train four days later, George had
somehow prevailed on Phil to assume a "spirit of graciousness,"
and a temporary détente was arranged. McCargar agreed to
reopen the lab and find funding for it somehow, but only if Phil
agreed to dismiss the newest, most trained, and higher-paid
members of the team. Varian and Brolly had to go. For "the
good of the cause," Phil accepted McCargar's terms.[45] But this
would not be the last time that Jess would demonstrate his
blatant disregard for the way Phil was trying to operate his
laboratory.

In the early months of 1931, the prospects for Television
Laboratories, Inc. took a turn for the better when 202 Green

Street received a visit from Walter Holland, vice president of engineering for a company commonly known as Philco.

"Philco" is a contraction of "Philadelphia Storage Battery Company." The company traced its roots back to the turn of the century, and it achieved some prominence in the early 1920s producing the batteries that powered most of the radios in use at the time. As household current became more prevalent, the need for battery-powered radios diminished and Philco turned its attention to manufacturing radio sets.

In 1930, Philco hit the mother lode when it introduced one of the most successful radio sets ever produced, the mock-gothic, deco-styled "Cathedral" Model 20. The Cathedral radio was so successful that Philco became the number one radio manufacturer in the country, surpassing even the previous market leader—RCA.

By 1931, Philco was selling more than twice as many radio sets as the so-called "Radio Corporation," a fact which was no doubt vexing to David Sarnoff and his minions as they saw their own recently acquired factory in Camden falling idle, while across the Delaware Philco's plants churned out thousands of units every day. The only consolation for Sarnoff had to have been RCA's ownership of all the patents. Even Philco could not manufacture radios without paying a license fee to RCA. Philco may have dominated the market for the time being, but only with the tacit blessing of RCA.

Conversely, those license fees were a constant source of irritation for the officers of Philco who had to sign the quarterly checks. It wasn't just the substantial sums involved that rankled the Philco bosses; it was the constant reminder of just who remained in command of the entire industry irrespective of who sold the most radios. And RCA had no reservations about reminding the industry just who was the final authority. The company's lawyers stated publicly, in hearings before the Federal Trade Commission, that RCA reserved the "right to sell or not to sell, to sell for a good reason or for a bad reason, or for no reason, and not to sell for a good reason, a bad reason, or no reason" licenses for the use of its patents.[46]

Such was the capricious, monopolistic embrace that RCA had over the entire radio industry. Nobody, least of all Philco, could manufacture radio sets or take them to market without the consent of the Radio Corporation of America. If the Radio Corporation withheld its consent by not renewing your patent license—well that was just tough, and you were out of business. Nobody understood this scenario better than the executives of Philco. Not even the size of their license fees earned Philco any special dispensation where the interests of RCA were concerned.

It was in this context that Mr. Holland of Philco, the largest seller of radio sets in the nation, visited the Television Laboratories in San Francisco to explore the possibility of licensing patents on a new art that RCA did not currently control. David Sarnoff was not the only influential figure in the industry who could look out to the horizon and see television as the tidal wave that would soon dislodge radio from America's living rooms. The executives of Philco surmised that, with their market clout and Phil Farnsworth's patents, Sarnoff's determination to dominate the new industry could be thwarted. By forming an association with Farnsworth, Philco hoped it could vault into a leadership role in the introduction of TV, and ultimately free itself from the oppressive yoke of its patent license obligations to RCA.

In the spring of 1931, Farnsworth ventured east again, this time to Philadelphia with McCargar, to finalize the arrangements for Farnsworth's first bona fide patent license. The details were rather complex, and negotiations dragged on for some weeks. While Phil and Jess were out of town, George Everson, back in San Francisco, was startled to receive an unexpected request for a visit to the Farnsworth labs: none other than David Sarnoff was in town and wanted to see Farnsworth's invention for himself.

Given the advanced state of the negotiations between Farnsworth and Philco—and the contentious relationship between Philco and RCA—it is not surprising that the most powerful executive in the communications industry would show up at an obscure loft in the warehouse district of San Francisco, three thousand miles away from his plush Manhattan offices. It is entirely possible that Sarnoff knew exactly where Farnsworth was that week, and what he was doing. As revealed in a later lawsuit, RCA operatives were not above enticing Philco secretaries with "expensive and lavish entertainment, hotels, and nightclubs [and providing] them with intoxicating liquors...in an endeavor to entice...said employees to divulge confidential information, data, designs, and documents."[47]

Furthermore, after Zworykin's visit to Green Street a year earlier, Sarnoff had been perfectly content to let the struggling Television Laboratories proceed with its undercapitalized research efforts, certain that anything of value could be acquired later.[48] But if Farnsworth was about to team up with RCA's biggest competitor, that might make a more formidable rival out of the upstart inventor and even preclude the possibility of an acquisition by RCA.

Sarnoff was also well aware by that time of the patents that had been issued to Farnsworth the previous summer, and realized he would be better off working with Farnsworth than against him. Unfortunately, the greatest obstacle to Sarnoff's working with Farnsworth—RCA's long-standing business practice of collecting patent royalties instead of paying them—remained in force. Herein lay the pivotal point of conflict between RCA and Farnsworth, who was not inclined to surrender his hard-won patents, as they comprised the cornerstone of his future enterprise. For Farnsworth, the commercialization of television was the means to an end; for Sarnoff, television was just another end.

Despite Farnsworth's apparent resolve, there was little doubt in Sarnoff's mind that eventually the Farnsworth patents would be brought under RCA's control. The only question for Sarnoff was whether RCA would be able to acquire the patents outright in a "friendly takeover" or would have to resort to the kind of grinding litigation that would strain the smaller company's finances so that it would ultimately capitulate. Sarnoff came to San Francisco to get a better idea of which scenario was most likely to unfold.

There was another reason why Sarnoff came to San Francisco. It was something more personal, beyond his bull-headed businessman's desire to head off any possible alliance between Farnsworth and Philco or assume control of Farnsworth's patents. Even this late in the game, Zworykin was still having trouble producing decent television pictures with tubes of his own design. So Sarnoff went to San Francisco to see something that he could not see anywhere else in the world—the state of the art in electronic television.

George arranged for Tobe Rutherford to demonstrate the system. Sarnoff, once inside the lab, looked all around to get his bearings, but as the system hummed to life, his gaze settled on the face of the receiver. Sarnoff studied the image with the chilling silence of a man who had confronted his own chosen future. It was startling enough that he was witnessing true television for the first time; what concerned Sarnoff more was that he had always expected to see it first under his own roof.

When he had seen all that he needed to see, Sarnoff drew Everson aside and quietly offered to buy the entire enterprise for an insultingly low-ball figure—$100,000. Sarnoff insisted, however, that the deal include the services of one Philo T. Farnsworth, who would have to come to work for RCA in Camden.

"Thank you for your offer," George replied to Sarnoff, "but Phil Farnsworth will never agree to such an arrangement. He's determined to maintain his independence any way he possibly can."

"Well, then," Sarnoff said, confidently dismissing the entire matter, "there's nothing here we'll need." With that, Sarnoff quickly departed, before George could ask him why he'd just offered $100,000 if he didn't really need it.

When he returned to New York, Sarnoff began pumping even more cash into Zworykin's operation in Camden. It might have been nice to own the Farnsworth patents. It might have been even nicer to have somebody as genuinely brilliant and clever as Philo T. Farnsworth working for RCA. But if that was not possible, Sarnoff was certain that liberal infusions of cash were all he would need to make up the difference between engineering and invention. The rest he would leave to the lawyers.

Negotiations between Farnsworth and Philco reached a successful conclusion in June of 1931 and Farnsworth finally had himself a patent license. Under the terms of the agreement, a major company agreed for the first time to pay royalties to the Farnsworth organization in exchange for the grant of rights to manufacture and sell television tubes and receivers. Clearly, Philco was anticipating that a healthy market would soon emerge for television products, and was ready to stake its claim in the new industry.

There were two "big wins" in the deal with Philco. For Phil Farnsworth, the most essential aspect of the agreement was its taking the form of a non-exclusive license. Farnsworth retained ownership of all his patents, the one point of business that he considered most vital to the long-range vision for his enterprise. Also significant was the term of the license, which would continue until August 26, 1947—guaranteeing Farnsworth a license for his first patents for the full life of their seventeen-year term.

For Jess McCargar and the other investors, the "big win" was Philco's willingness to assume the cost of Farnsworth's ongoing research operations. The cost of his further research would be deducted from future royalty payments, but at least McCargar and Everson would be freed from the endless obligation of raising more funds.[49] The only catch was that Philco wanted Farnsworth to move his entire operation from San Francisco to Philadelphia, to set Philco up in the business of television.

Phil, flush with the excitement of having his first big license, was looking forward to the move. Not only would he be out from under McCargar's unsympathetic thumb, he would also be closer to the center of the action on the East Coast. Laboratories like Bell Labs and RCA were beginning to realize the importance of television and invest in its future, and the Federal Radio Commission in Washington was beginning to consider the implications of the new medium and its need for spectrum allocation and signal standards. Phil wanted to be close to all of those debates.

Pem was considerably less enthusiastic. Not even two years had passed since the family had moved into their lovely new home in the Marina district of San Francisco and she hated the thought of giving that up. She had given birth to the couple's second son, Kenny, in January, and she was not looking forward to traveling cross-country with a nursing infant and a toddler in tow. Phil tried to reassure her that the move was only temporary, that they'd be back in their home by the Bay in six months.

Word of the impending move made it all the way back to Utah. *The Salt Lake Tribune* captured the significance of the relocation in a headline that trumpeted the imminent arrival of commercial television: "Utah Youth Goes East to Build Production Model." Clearly, the race for television was firing on all cylinders, and Philo T. Farnsworth was leading the pack.[50]

Packing the lab turned out to be a much more daunting task than anyone had bargained for. Nobody realized just how much gear had accumulated at 202 Green Street until they tried to crate it all up to put on a freight car. The Farnsworth family, and the lab gang and their families, took over one whole passenger car of the eastbound steam train. The trip began in the middle of July, and as the train headed east, it left behind the cool breezes of San Francisco, chugged over the High Sierra, and rattled into the Plains States in the midst of a stultifying heat wave.

"The train was so hot that the thermometer in the dining car reached 140 degrees and exploded," according to Pem. "Of course, there was no air conditioning. We had a whole car, and spent most of the days in the drawing room. Our baby was only about five months. We just stripped him down to his diapers, and put him on the bed, and little Philo was stripped down to little shorts, and we let him run with the other kids...."[51]

The heat wave persisted when the entourage reached Philadelphia a few days later. The climate wasn't the only thing different. The facilities at Philco were not at all like the familiar homespun atmosphere of the loft in San Francisco, and the delicate necessities of life under the wing of a large corporation presented quite a change for the Farnsworth lab gang. Among the starch-collared, book-educated Philco engineers, Phil and his boys were regarded as mavericks, a gang of crazy cowboys from California. Patience often wore thin, particularly when the intense summer heat turned Farnsworth's uninsulated, top-floor lab into a virtual oven. On one extremely uncomfortable day, Cliff and his crew in the tube lab abandoned not only protocol, not only their coats and vests and ties, but their shirts as well—a tonsorial departure that prompted one well-heeled Philco executive to call Farnsworth and his men "animals."

Not only was the work environment a difficult departure from the lab in San Francisco, but the managers of the operation for Philco insisted that the work itself be conducted in an atmosphere of absolute secrecy. According to George Everson, "No one except Mr. Holland, the vice president in charge of engineering (who had been instrumental in advocating the license with Farnsworth), Mr. Grimditch, under whose department the laboratory functioned, and Mr. Skinner, the president, were allowed to visit Phil's sanctum." George speculates that "the policy of secrecy was designed to prevent annoyance and interruption by the curious,"[52] but George was always charitable when it came to describing the internecine undercurrents that simmered beneath Phil's work. It seems equally likely that Philco was not anxious for the big company just across the river in Camden to find out what was going on atop the old brick factory at Ontario and C Streets in Philadelphia.

However, the veil of secrecy became impossible to maintain when Philco obtained an experimental license from the FCC to conduct over-the-air television transmissions. The company set up a transmitting antenna on the roof of the building that housed the laboratory, and Philadelphia's first television station, W3XE, was on the air.[53]

Farnsworth set up a prototype receiver in his home, where two-year-old Philo III and his baby brother Kenny became the first charter members of the "television generation." Their usual program diet consisted of the Mickey Mouse cartoon, "Steamboat Willy," which ran over and over again through the film-chain at the laboratory several miles away. While the little Farnsworths watched, their father and the engineers at Philco made adjustments and tuned the circuits.

Philco and Farnsworth were not the only researchers in the area who were now testing television broadcasts. RCA also had an experimental license, and the Farnsworth engineers began picking up the RCA test transmissions at the Philco plant. From the pictures they were picking up in Philadelphia, Farnsworth and his crew made an important discovery: In the first months of 1932, it appeared that Zworykin was testing a new kind of electronic camera tube—something other than an ersatz Image Dissector. The pictures were not nearly as sharp as those Farnsworth was generating with his own system, but they did cause him to wonder if there was something to Sarnoff's boast in San Francisco that "there's nothing here we'll need."

Despite the accelerating competition with RCA, one blessing for Farnsworth emerged amid all the complexities of working within a traditionalist company like Philco. With much of the television work now in the hands of his assistants, and with the continuity of their employment assured, at least for the time being, Phil found that he could begin turning some of his attention to the more advanced, abstract, and compelling ideas that had caught his attention while he was experimenting in San Francisco.

This mental elbow room let him seize the opportunity to investigate more closely the phenomenon of secondary emissions which he had observed during his experiments with amplification. Despite making substantial improvements over the previous four years, Farnsworth was still having problems. with the sensitivity of his Image Dissector tubes, and he continued to believe a solution was to be found in the secondary emissions of electrons. At one point, he actually implanted a small electron multiplier in the neck of the Image Dissector tube, which caused a tremendous jump in the strength of the signal coming out of the tube. With that success, he began to contemplate an even more impressive application for the secondary emissions effect, a kind of "super amplifier" tube that would produce power levels far beyond anything previously achieved.

However, everything going on at Philco—conflicts with management, successes in the lab, the designs for a new tube—became totally insignificant when a personal tragedy struck the Farnsworth family in the winter of 1932. A virulent streptococcus infection struck Phil and Pem's second son, "our sweet little Kenny of the blond curly hair, large blue eyes and laughing disposition."[54] The illness severely congested the baby's throat and restricted his breathing.

His parents rushed the baby to the University Hospital in Philadelphia. "We all felt so helpless," Pem recalled. "There was nothing we could do for him, and there were no sulfa drugs, no penicillin, no streptomycin. The only thing the doctors could do was perform an emergency tracheotomy." A breathing tube was inserted into the child's throat, rendering him speechless. "It tore at our hearts to see him so sick and so frightened at being unable to make a sound." But even that drastic measure was insufficient to loosen the grip of the infection.

A pediatrician stayed with the child into the night, until he seemed to be out of danger. He left an intern in attendance and encouraged the parents to get some rest in an empty hospital room nearby. Pem nodded off, but Phil could not sleep. Shortly after midnight, Phil went to look in on the baby, only to find that the intern had fallen asleep—and Kenny was turning blue. The doctors tried to revive him, but "we had to stand by helplessly and watch his precious life slip away. Suddenly, unbelievably, our darling son was gone."[55]

As tragic as the loss of their child was, nothing could prepare the grieving parents for the cold measure of insensitivity with which the bosses at Philco greeted the news. Phil and Pem wanted to take Kenny back to Utah to bury him among the other dearly departed souls of the Farnsworth and Gardner clans. But the Philco people would not permit it. Phil was informed in no uncertain terms, "We just can't spare you at this time" and forbidden to accompany his wife for the heartbreaking trip back to Utah.

Pem recalled later, "I don't know why he didn't just say 'well, that's too bad...' and come with me." What she could not have realized was the chilling effect that Phil's grief was having on him, or the devastating impact it would soon have on their relationship.

So Pem was forced to make the tearful trip to Utah alone, later writing, "It is hard to relate the agonizing loneliness and grief that I suffered during those four seemingly endless days crossing the continent, knowing my baby lay cold and alone in a coffin in the baggage car ahead."[56]

Difficult as that trip west was, life became even more trying when Pem returned to Philadelphia ten days later and discovered that her husband was hardly speaking to her.

NINE

Suspended Animation

Farnsworth with his camera at the Franklin Institute

"Even if you're on the right track, you'll get run over
if you just sit there."
—Will Rogers

When the Farnsworth family moved to Philadelphia, Phil assured Pem that the situation was only temporary, that they would get Philco set up in the television business and be back in their cozy "little home in San Francisco" in six months. As six months turned into two years, it was clear to everyone involved that the Farnsworth operation would not be returning to California anytime soon, if ever. In the meantime, relationships seemed to be deteriorating on all fronts. The atmosphere around the Farnsworth lab at Philco became just as chilly as the frosty silence that formed on the Farnsworth home front in the wake of little Kenny's death and Pem's lonely trip to Utah.

The secrecy surrounding the television operations at Philco contributed to an air of resentment that simmered among other Philco engineers. It wasn't just the sartorial differences—the absence of neckties or the rolled-up sleeves. There was something about an inventor working in a shop of non-inventors that seemed to breed its own measure of contempt, as if the inventors were somehow smarter than the other engineers, or less beholden to the corporate masters who dictated the day-to-day activities of the other engineers.

About this time, Phil began having suspicions of his own. He began to sense that something was going on behind the scenes, that something unsavory was brewing between Philco and Jess McCargar. It occurred to him that maybe McCargar was negotiating an outright sellout to Philco.

Farnsworth had never really trusted McCargar to protect what he considered to be his long-range interests. He had always deferred to George Everson's accommodating intervention whenever issues with McCargar boiled over. Now, however, Phil's misgivings became so intense that he felt compelled to stop keeping his daily laboratory journals, so that the ownership of any new ideas could not be challenged. This change in his regular routine forced him to work out the intricate details of his new "super amplifier" tube without the aid of any written notes or equations. This was an extraordinary measure, because Farnsworth was well aware of how precise laboratory records could serve as a critical form of documentation in the event his patents were ever contested. But he had to weigh the risk of inadequate records against the possibility that Philco might one day claim ownership of any new inventions that he disclosed while he was working under their roof. Given the deteriorating circumstances around the Philco plant, he decided that keeping his new ideas to himself posed the lesser risk.

This cloak-and-dagger atmosphere had made Phil an infrequent correspondent with Jess McCargar and George Everson, who remained on the West Coast; now he stopped corresponding at all. For nearly three months, neither Jess nor George heard a word from Phil, until one day in the spring of 1933 when Jess accepted a collect call from Philadelphia.

"Jess," Phil told his primary benefactor, "I've made a decision. I've left Philco. I've taken what equipment I could here to my house, and I'm going to set up my own lab."

McCargar responded in a seething rage. "Well then, I guess you'll be coming back to San Francisco."

"No, Jess, I'm not," Phil answered. "There are too many important things happening here on the East Coast. This is where all the action is now, and we need to be close to it. The government will be considering spectrum allocation for television broadcasting soon, and I want to be close to those deliberations. We have to stay here in Philadelphia."

McCargar slumped as he considered the implications of Phil's announcement. If Farnsworth had indeed walked out on Philco, then Television Laboratories, Inc. could no longer count on Philco to fund the ongoing research and refinement operations. McCargar was less than thrilled by the prospects of peddling stock in what was still a highly speculative venture in the midst of the Great Depression. Perhaps McCargar thought at this point that he could discourage Farnsworth by simply refusing to raise any funds, threatening to cut him out altogether.

"Where are you gonna get the money?" McCargar growled defiantly over the wire.

Once again, Phil summoned his poker instincts and called Jess's bluff. "If you can't find the money, then I will," Phil answered firmly and hung up.

No sooner did the line go dead than Jess and George piled onto the transcontinental express and headed for Philadelphia. By the time the two businessmen arrived, Farnsworth had already removed what equipment he could from Philco's facility and was looking for a new place to set up shop. In the meantime, the equipment was scattered about the Farnsworth living room, where the three partners assembled for the first time in nearly two years.

Once the emotions were all played out and the discussion settled down to business, McCargar agreed to resume raising operating funds so that the job of perfecting Phil's invention and making it suitable for the marketplace could proceed. Phil reluctantly accepted the concessions that McCargar demanded, the most painful of which required Phil to pare down his staff again. That meant letting go some of the men who had made the trip from San Francisco, including Harry Lyman and Russ Varian (who had agreed to return to the fold after being fired the first time McCargar shut off funds).

Jesse B. McCargar

Phil hated to let those two down again. Some of the men agreed to find other work in Philadelphia until Phil got back on his feet and could hire them again. But Russ Varian and others, as much as they revered Farnsworth and valued working with him, decided they'd had enough of Jess McCargar's management of the operations, so they returned to San Francisco.[57]

Yet another incarnation of the Farnsworth venture was incorporated, with the name of Farnsworth Television, Inc. Phil found a suitable location at 127 East Mermaid Lane in a suburban neighborhood near Philadelphia, and with the underpaid help of Cliff Gardner and Tobe Rutherford, began rebuilding. Their task was formidable. Most of the important equipment that they needed for their work was the property of Philco and had to be left behind. They were building from scratch again.

This time the system that Farnsworth was building was a far cry from the crude wooden boxes he'd built back in the days of 202 Green Street. The whole system had an air of sophistication about it. The camera units were no longer so imposing that they had to be bolted to a work bench; they were compact enough to be mounted on a tripod, like a movie camera. The viewing screens were displaying images with more than 300 lines per frame on screens seven to ten inches in diameter. Indeed, the progress of the past six years was demonstrable. Television looked like it was ready for public consumption.

So equipped, Farnsworth and his supporters resumed their effort to find another company that would take a license for the ever-growing Farnsworth patent portfolio and thereby join them in their quest to deliver television to the marketplace. Much to their consternation, there were no takers.

Through contacts in the industry, Farnsworth and his backers learned why none of the most likely candidates would offer Farnsworth a license for his patents. These companies were all actively engaged in the manufacture of radio equipment and were dependent on patent licenses with the Radio Corporation of America for their very existence. Farnsworth's people learned that RCA had issued an unwritten edict to their licensees: work with Farnsworth, and their radio patent licenses might not be renewed.

By this time—mid-1933—RCA was conducting experiments and demonstrations with an improved version of the camera tube that Farnsworth first observed while monitoring test transmissions from the RCA lab in Camden. Sarnoff and Zworykin finally had succeeded in contriving an electronic television system comparable to Farnsworth's—more than three years after Zworykin's visit to Farnsworth's lab.

The device that Zworykin was working with was an odd duck of a camera tube he called the "Iconoscope." Though the result was more or less the same—an electronic television picture—the Iconoscope was very different from the Image Dissector. Farnsworth's tube featured a simple, straightforward design, which at least one observer described as having "everything that it needs and nothing that it doesn't." In contrast to the elegant design of the Image Dissector, the Iconoscope featured an awkward, triangular geometry in which the light from the image source and the scanning electron beam were each focused on the same side of a photoelectric surface but from different angles. This geometry made accurate, consistent scanning of the image difficult.

Furthermore, the photoelectric surface of the Iconoscope was made up of tiny, individual islands of photosensitive material that were capable of "storing up" the electrical charge that accumulated in the interval between scans. Using this "storage principle," the Iconoscope could generate a stronger signal than the Image Dissector, but any movement of the subject during the interval between scans would cause a variety of values to be stored before the current was discharged, producing a muddled picture. Because the Image Dissector used a continuous surface and delivered only the electrical value at the precise instant of scanning, the Image Dissector produced a sharper image even though its signal was weaker than the output from the Iconoscope.

Clearly, the Iconoscope that Zworykin began demonstrating in the early 1930s was a very different tube from the Image Dissector that Philo Farnsworth first demonstrated in the late 1920s. Nevertheless, once RCA had the Iconoscope working and producing an image comparable to Farnsworth's, RCA began a campaign to declare that the Iconoscope and the Image Dissector performed similar functions. More significantly, RCA went on to claim that the Iconoscope was the very same tube that Zworykin had disclosed when he applied for a patent for an electronic television system way back in 1923, while he was working at Westinghouse. Presumably, this was the same system his superiors at the time had dismissed, telling him to "work on something more useful."

The patent-related claim was critical. If indeed the Iconoscope and the Image Dissector performed a similar function in a similar manner, and if the Iconoscope was indeed the tube that Zworykin disclosed in 1923, then RCA could make a case that Zworykin's invention predated Farnsworth's—and that anybody licensing the Farnsworth patents would be infringing on whatever patents Zworykin eventually received for the Iconoscope.

In press releases generated at the time, RCA's praise of Zworykin's contribution was extensive, although just what Zworykin contributed that made the Iconoscope uniquely his invention is unclear. The storage principle that represented such an important element of the tube's design actually owes its origins to work conducted in the 1920s by a Hungarian scientist named Kolomon Tihany.[58] It is also worth noting that at the same time the

Iconoscope appeared in America, an identical device, called the "Emitron," was introduced in England, where its invention was attributed to J.D. McGee and others working for the British conglomerate EMI.

EMI enjoyed a patent cross-license with RCA, so it is hardly coincidental that the Iconoscope and the Emitron would surface in two places at the same time. What is less clear is where the tube really appeared first. Regardless of its uncertain origins, the Iconoscope was nevertheless a device upon which David Sarnoff was determined to base the extension of his empire and go down in history as the man who gave television to the world.

Once the Iconoscope was working satisfactorily, Sarnoff began quietly spreading the gospel among RCA's licensees that doing business with Farnsworth was off-limits. That was all RCA needed to do to put Farnsworth's search for a new patent licensee out of reach. Because Farnsworth had been unwilling to settle when David Sarnoff came to San Francisco and offered the trifling sum of $100,000 for the entire operation, Sarnoff was putting "plan B" into motion.

In the decade since RCA was first formed around the earliest radio patents, the company had built an imposing track record of reducing erstwhile competitors to corporate rubble through an unbroken string of successful patent litigations. Now Sarnoff directed his legal battalions to begin maneuvering to bring Farnsworth's portfolio of patents under RCA's control.

But Sarnoff underestimated the determination of Philo T. Farnsworth. Realizing the bind they were in, Farnsworth's team took the offensive and did the only thing they could do: they mounted their own challenge of RCA's claims before the examiners of the U.S. Patent Office.

Sarnoff must have salivated at the news that Farnsworth was challenging RCA. The case could not have come at a better time. In his relentless pursuit of an alternative technology that could circumvent Farnsworth's patents, RCA was spending money on television research at *ten times* the rate that Farnsworth was—all at a time when the economy was still suffering the effects of a crippling depression. Sarnoff needed a quick return on all that investment in order to preserve his stature and calm the rumblings within his board of directors. To achieve his purposes, and to continue RCA's practice of collecting patent royalties into the new art of television, Sarnoff needed a clear path on the patent front. And here was this upstart, Farnsworth, throwing himself in front of the steamroller.

Farnsworth's attorney, Donald Lippincott, drew careful aim with his pencil and legal pad. He knew exactly where to strike and how to nullify RCA's claims for Zworykin's Iconoscope. He focused the interference proceedings on Claim 15 of Farnsworth's 1930 patent, #1,773,980, which describes the simple, elegant concept of an "electrical image"—the vital step in the process of converting light into electricity. Something slightly intangible is embodied in the precise wording of Claim 15 that reveals the indispensable process of creating an electrical counterpart of an optical image, in which values of electricity correspond to values of light. Claim 15 describes:

> An apparatus for television which comprises means for forming an electrical image, and means for scanning each elementary area of the electrical image, and means for producing a train of electrical energy in accordance with the intensity of the elementary area of the electrical image being scanned.

This paragraph, first composed by Donald Lippincott in 1927, announces the arrival of television. The words express essentially the idea that fourteen-year-old Philo T. Farnsworth pictured in his mind's eye that hot afternoon while crisscrossing the fields on his disc harrow near Rigby, Idaho. Claim 15 describes the essence of Farnsworth's invention, the missing ingredient, which, once found, paved the way for television as we know it. Yet, in the 1930s, RCA was trying to claim that Zworykin had the same idea first.

A patent interference case is not ordinarily a courtroom drama with witnesses and lawyers and dramatic unscripted testimony.

Rather, the proceedings are carried out through lengthy depositions and briefs, which are submitted to the patent examiners for
their written ruling. Nevertheless, the burden of defending his
patents weighed heavily on Farnsworth personally. RCA's attorneys descended on Farnsworth's office, and Donald Lippincott
flew in from San Francisco to be present for the depositions. As
Pem described it, "They came to Phil's office and held court there
for two or three weeks. When they left, Sam Smith (one of the
RCA attorneys) apologized to Phil for putting him through all
that, saying, 'You know, it's my job.' They went on from one
place to another, taking depositions from other people."[59]

Reams of testimony were taken. Every stack of depositions
meant another week that Farnsworth was kept out of his laboratory, another week of progress lost to the competition. Despite
the distractions, Farnsworth and Lippincott saw right through
RCA's semantic charades and chipped away at RCA's case. They
built clear, concise, and uncompromising arguments that me-
thodically demolished RCA's claims.

Much of RCA's case focused on Farnsworth's testimony that
he had first thought of his approach to electronic television as a
high school freshman. RCA's attorneys greeted this contention
with a derisive laugh—how could a mere child possibly dream
up something as elaborate as electronic television? Farnsworth's
story stood in stark contrast to Zworykin's extensive credentials,
his education, and his years of service with companies such as
Westinghouse and RCA. Certain that Farnsworth couldn't possibly
substantiate the claim, the opposition pressed the point, and challenged Farnsworth and his attorneys to produce the one person
Farnsworth told about his ideas—his high school chemistry
teacher.

What genuine drama such proceedings can deliver unfolded
when Samuel Smith, representing RCA, and Donald Lippincott,
representing Philo T. Farnsworth, succeeded in locating Justin
Tolman in Salt Lake City. They found him one afternoon outside
his house, working in his rose garden.

As Pem recounted, Lippincott asked Tolman, "Do you re-member a student in Rigby, Idaho by the name of Philo Farns-worth?" To which Tolman replied amiably, "I surely do...he was the brightest student I ever had."[60]

Not wanting to prime the witness, Lippincott didn't ask any more questions. He asked Tolman if he would give them a deposition, and Tolman agreed to meet the lawyers the next morning at an office in the city. Lippincott worried all night over what Tolman would say, and what he remembered, but the attor-ney's apprehensions turned out to be totally unfounded. Tolman remembered Farnsworth quite well, and told the two attorneys that he had even bragged about his prodigy to his subsequent classes. He described the conversation that he and his pupil had had after school, the drawings the boy sketched on the black-board, and what he remembered of the television system that Philo had described. And then Tolman delivered the *coup de grace*: He "took from his pocket a well-worn sheet torn from a small pocket notebook whereon Phil had sketched his Image Dis-sector tube, saying 'This was made for me by Philo early in 1922.'"[61] Lippincott took the drawing from Tolman, smiled, and handed it to Sam Smith. The RCA attorney just shook his head in silence and handed the drawing back to Lippincott.

Perhaps RCA wanted to undermine the legitimacy of Farns-worth's claim because their own case was so surprisingly weak. There was no effort to produce into evidence a tube from 1923 that would substantiate Zworykin's claim to have had an oper-able television transmitter at that time. There were some vague verbal accounts, but those were dismissed by the examiners as unreliable, having been "influ-enced by later events and knowl-edge." In other words, when it mattered most, RCA was either unwilling or unable to produce the evidence that would support

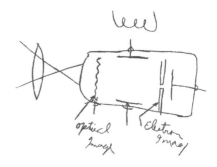

Justin Tolman's notepad sketch, hand-drawn by Philo in 1922

Zworykin's claim to have invented the Iconoscope—or something like it—in 1923.

In February of 1935, the U.S. Patent Office delivered its milestone decision in the case of *Zworykin vs. Farnsworth*. In its final ruling in case number 64,027, the patent examiners summarily dismissed the larger corporation's claims in terms that spoke almost derisively of RCA's entire presentation, saying in conclusion that:

1. Zworykin has no right to make the count by virtue of the specific definition of the term "electrical image" given in the Farnsworth patent;

2. Zworykin has no right to make the count because it is not apparent that the device would operate to produce a scanned electrical image unless it has discrete globules capable of producing discrete space charges and the Zworykin application as filed does not disclose such a device;[62]

3. Zworykin has no right to make the count even if the device originally disclosed operates in the manner now alleged by Zworykin because this alleged mode of operation does not produce an electrical image that is scanned to produce the television signals.

After a few more pages of legal discourse, the decision ends with an unequivocal declaration:

"Priority of invention is awarded Philo T. Farnsworth."

Unfortunately, this resounding proclamation was followed by one more little sentence: "Limit of Appeal: August 22, 1935." In other words, RCA could still appeal the case; the company waited the full six months to file their appeal, which in turn was not heard until January 1936. The patent office took another two months to consider the appeal, and in March of 1936 upheld the original decision. After the initial appeal was denied, RCA's only recourse was to take the case to a civil court—and it had another six months to file that appeal.[63]

With little money left to carry on the fight, Farnsworth and his supporters held their collective breath, waiting to see if RCA would extend the ordeal by appealing the civil case. RCA took every day of those six months before it decided not to appeal. This news was welcomed by the beleaguered Farnsworth camp, but there was little cause for celebration, for the pattern was clearly drawn: Farnsworth's entanglements with RCA would go on for years, and placed the future of television in a perpetual state of suspended animation.

There was little solace for Farnsworth in his victories over RCA—rare as those were in the annals of electronics history. The giant company continued to do everything in its power to avoid coming to terms with Farnsworth—terms that would have required RCA to deviate from its long-standing, oft-stated philosophy that it was in business "to collect patent royalties, not pay them." David Sarnoff shrugged off the patent examiners' decision and remained as determined as ever to deliver television to the marketplace on his own terms, using technology developed in his own laboratories, with timing that suited his own corporate priorities—and which coincided with the expiration of the patents already under his control. The decision in the 1934 interference, with its lengthy trail of appeals, was just the first in a string of Phyrric victories, each a successful battle for Farnsworth in a war that seemed no closer to ending.

Things remained equally bleak on the home front. After McCargar agreed to resume funding for the laboratory operations following Phil's departure from Philco, Phil's salary was drastically cut, and he was forced to relocate his family to a small, unfurnished house in Philadelphia. He, Pem, and their remaining child, Philo III, moved into it with "only beds and two chairs." Were it not for a breakfast nook in the kitchen, the family would have been reduced to eating off the floor.

The bareness of the surroundings seemed to reflect the pain the family was still feeling in the wake of little Kenny's death. Pem had long since given up whatever hope she had left that they would ever return to their little "love nest" by the Bay. Now she was even beginning to despair of even preserving whatever was left of her relationship with Phil. As Pem put it, ". . . the barrier between Phil and me grew daily as we bottled up our emotions following Kenny's death...we lived together almost as strangers."[64]

Phil and Pem continued to "go through the motions" of being a couple despite the cold distance between them. Among the activities they continued was their practice of attending dances on Saturday nights. Pem noticed that Phil was beginning to actively flirt with other women, and was most frequently taking to the dance floor with a married woman named Kay—with little regard for the impact these dalliances had on Pem's feelings. Pem decided two could play that game, and began her own flirtation with Kay's husband, a "tall, flamboyantly handsome blond fellow" named Hank. Phil and Pem hardly danced together at all, and Pem found herself fantasizing about "running off to Reno for a divorce" so she could marry Hank.[65]

One night, the orchestra began to play the tune that Phil and Pem had always considered their song, Irving Berlin's "Always." The singer crooned,

Days may not be fair Always,
That's when I'll be there Always.
Not for just an hour,
Not for just a day,
Not for just a year,
But Always.[66]

Suddenly Pem looked up and found Phil standing before her, sheepishly saying, "I think this is our dance." Pem didn't know how to answer. She just put her hand out, and Phil escorted her to the dance floor.

"At first it was like dancing with a stranger," Pem recalled. "Then he held me close and whispered in my ear, 'What's the matter with us? We must be crazy! Let's get out of here, we need to talk.' We went home and talked the whole night through."[67]

After months of virtual silence between them, Pem and Phil finally discovered what had been driving them apart. Phil, the scientist, the inventor, the magician...felt he should have been able to do *something* to prevent Kenny's death, and in his failure to produce a miracle came to believe that Pem thought it was *his* fault. Likewise, Pem had been overcome by her own sense of helplessness, and thought there must have been something that she, the child's mother, could have done to avert the tragedy of his death—and so had convinced herself that Phil thought it was all *her* fault. In their isolated grief, each thought the other held them accountable.

With this revelation, Phil and Pem began the long and painful path of sharing their grief, recovering their devotion, and reconciling their relationship. "From that night on," Pem said, "we gradually rebuilt our relationship, this time with such a close bond that nothing ever threatened us again."[68]

Though the patents remained mired in litigation for years, the business side of the Farnsworth's lives took a favorable turn in the summer of 1934, when the prestigious Franklin Institute of Philadelphia invited Philo T. Farnsworth to conduct the world's first full-scale public demonstration of television.

Encouraged as he was by the vastly improved performance of his entire system, Farnsworth accepted the invitation, disregarding for the time being his stalemate with RCA. After all, the future of television belonged not with any single corporation, but with the people, the audience that would buy television sets and watch television programs. Farnsworth hoped to score some points with the public by being the first to show them what they could expect.

While Farnsworth was preparing for the Franklin Institute exhibit, he was introduced to Russell Seymour Turner, known to his friends as "Skee." Turner was an engineer and businessman whose wealthy father had acquired a substantial chunk of Farnsworth stock, which George Everson and Jess McCargar continued to sell to raise funds for the ongoing research. Skee was sent in to see what he could do to push the enterprise closer to some sort of commercial payoff. Skee was smitten immediately with the

Farnsworth charm—what Pem often called "the Farnsworth effect"—and began to take a strong personal interest in Phil and what he had to offer.

Russell Seymour Turner

Turner saw to it that Farnsworth had enough funds to build a completely new system for the Franklin Institute exhibit. As a result the picture tube that Cliff made was the size of a ten-gallon jug, and the camera was compact even by today's standards. So equipped, Farnsworth was handsomely prepared to introduce his invention to the populace that he hoped would soon embrace it.

The exhibit was an unprecedented success. There was little advance publicity—only word-of-mouth—but people were lined up for blocks when the doors opened in August. The response was so strong that the event, originally scheduled to last ten days, went on day and night for three weeks.

Farnsworth placed one camera unit near the door, and the power of his invention was instantly driven home to anyone who entered, as they were immediately confronted by their own disembodied image flickering across the bottom of a ten-gallon bottle.

In addition, programs were thrown together spontaneously and transmitted from the roof of the Institute to an auditorium downstairs. Vaudeville acts, popular athletes, and a swarm of politicians volunteered to appear before Farnsworth's cameras. Thousands of Philadelphians poured through the auditorium in fifteen-minute intervals to see whatever was appearing. The crowds were indifferent to the content. They came to see the image on the screen, whatever it was. They came to witness the ancient dream of seeing at a distance. For the depression-weary populace, this was something really new—something that spoke of a future, a visible, tangible oracle of better times to come.

For Farnsworth and his men, success at the Franklin Institute was a terrific morale booster. It was their first contact with so

large an audience; their first undeniable proof of how big television was going to be. But there was another side to the Franklin
Institute event, which is apparent only in photographs taken during the event. These pictures show a very different Philo Farnsworth from the earlier photos of the boy inventor and his magic
fruit jars. This Philo Farnsworth, at the age of twenty-eight, looks
thin, drawn. One picture in particular shows Phil standing next to
his nifty new compact camera, his brow furrowed, his eyes
squinting, his face gaunt. Simply put, the man looks weary.

Considering all the forces that had converged on this one
man, it's no surprise that the strain was beginning to show. It is
not enough that he was tackling the unseen forces of nature
every day in his laboratory; he had to face these challenges in an
increasingly uncertain environment. Not only was he dealing with
fierce competition from a formidable corporation that wanted to
deny his accomplishments and declare them its own, but he was
also constantly shackled by the unwillingness of his own financial
sponsors to share his long-range vision. Television may have
been an oracle of better things to come, but for Phil Farnsworth
they could not come soon enough.

The Franklin Institute demonstration also attracted considerable international attention and marked the beginning of a
steady flow of foreign visitors to Farnsworth's lab at 127 East
Mermaid Lane in the Philadelphia suburbs. Scientists and digni-

Farnsworth aims his camera on the roof of the Franklin Institute

taries from all over the world came to see the miracle in Farnsworth's living room. Phil and Skee Turner learned a great deal from their guests about the state of television around the world. They were particularly interested in stories from England, where the BBC had been conducting experimental video broadcasts for some years.

The system that the BBC was using was a mechanically scanned device that was

Phil checks out the receiver

invented by the eccentric Scotsman John Logie Baird. After dabbling in a variety of businesses—socks, jams, glass razors, and soaps—Baird had turned his attention to television, constructing a system based on the Nipkow disk. Baird achieved some notoriety in England in 1926, beginning when he transmitted a crude image of the head of a dummy from one room to another. He followed that landmark achievement with a successful transmission of blurry but recognizable television images by radio across the Atlantic. With these milestones under his belt, Baird convinced a reluctant BBC to permit him to use its broadcast channels in the evenings to broadcast blurry programs to a handful of receivers. By 1934, Baird had sold more than twenty thousand Televisor receivers in kit form all over Europe. Still, the BBC was disappointed in the quality of Baird's picture and started looking for something better.

As providence would have it, Baird's people learned of an inventor in America who was looking to license an electronic television system, and quickly dispatched a group of engineers to Philadelphia to see what the young man had to offer.

Phil Farnsworth and Skee Turner received the news of a possible license from Baird with tremendous excitement—a license in England could be the prelude to a whole series of licenses all over Europe. On arriving in Philadelphia, the English engineers were instantly impressed with Farnsworth's system, and at their invitation arrangements were made to take Farnsworth and his invention to England, where negotiations for a license would be concluded.

Phil and Skee Turner reveled in the unexpected change of fortune. Despite the ongoing litigation with RCA, there was new hope for television across the Atlantic. So Philo T. Farnsworth carefully crated up his circa-1934 television "mobile unit" and sailed for Southampton, hoping to accomplish in Europe what he could not accomplish in America.

TEN

We Want Cash!

Philo Farnsworth, Tobe Rutherford and
Skee Turner aboard the S.S. *Bremen*

"When a true genius appears in this world you may
know him by this sign, that the dunces are all in
confederacy against him."
—Jonathan Swift

In the dark and starry nights he spent on the deck of the S.S.
Bremen, Philo T. Farnsworth, still not thirty years old, had ample time to reflect upon the unlikely chain of events that put him
aboard a luxury liner steaming from New York to England.

By the fall of 1934 Philo's daring invention of electronic
television should have placed him at the vanguard of the communications industry. However, the world's largest radio manufacturing and broadcasting concern, the Radio Corporation of
America, was committing a formidable amount of manpower and
money to fight the tiny Farnsworth Television Company over the
fundamental patents for electronic television. For Philo, his lab
gang, and the scattering of patient investors his enterprise had

attracted over the years, the ongoing litigation with RCA had a twin-edged effect. As long as Farnsworth's patents remained under contention, his company was forestalled in its effort to grant licenses and collect royalties on his inventions. Yet without such funds, Farnsworth was hard-pressed to maintain the legal effort necessary to defend those patents, or meet the rising cost of the research necessary to maintain his lead in the field.

Nor did it matter that the patents had a limited, seventeen-year term. The patent law made no provision for the fact that a competitor was steering business away from Farnsworth's patents. Those patents were legally in effect, and the clock was ticking.

Pressure was building from some backers who would have had Farnsworth accept RCA chief David Sarnoff's terms, which amounted to a total sellout of the patents to RCA. Worse for Farnsworth, he would have been forced to become a Sarnoff employee—a condition he considered only a small step up from indentured servitude. So the legal entanglements continued, as RCA challenged other elements of the Farnsworth patent portfolio and then appealed every decision the patent office rendered in Farnsworth's favor.

One such case, around the "blacker than black" scanning technique (patent #2,246,625) stretched on for a full eleven years.[69] Consequently, funds that should have been used to make Farnsworth's invention ready for the marketplace—or, better yet, to branch into entirely new areas of scientific inquiry—were siphoned off into the ongoing legal battles.

As he sailed for Europe, Farnsworth knew that his portfolio contained many patents that were unavoidably essential to the art of turning light into electricity by means of a pencil-thin, rapidly deflected electron beam. Subsequent developments by other companies, including those produced by RCA, proved that this was indeed the direction that the rest of the industry would follow. But owning the patents was not in itself sufficient to guarantee Farnsworth's personal or professional prosperity.

Since domestic markets were forestalled indefinitely by RCA's dominance and the cloud of litigation, Farnsworth had no choice but to seek foreign alternatives for licensing his patents. When Baird Television invited him to bring his invention to England to be considered for a patent license there, Farnsworth was certain he had found a timely solution to the costly delays at home.

Whatever success Baird had achieved with his mechanical television system was largely due to his tenuous relationship with the British Broadcasting Corporation, which controlled virtually all broadcasting in the country. Though never terribly keen on Baird's system, during the early 1930s "the Beeb" permitted Baird to use their radio channels late at night to broadcast pictures on a temporary, experimental basis. Using one radio channel for his low-frequency, low-resolution pictures, and another channel for sound, the small cadre of radio amateurs who assembled his receiver kits were rewarded for their diligence with a fuzzy preview of the age-old dream of seeing from a distance.

Unfortunately for Baird, the costs of tooling up for production forced him to seek financial assistance, and in the process he lost control of his company to a large conglomerate called British Gaumont. This arrangement worked fine for Baird until 1934, when the BBC was shown a version of electronic television that had been developed in the laboratories of the EMI Corporation (formerly Electrical and Music Industries, Ltd., and sometimes referred to as "Marconi-EMI" owing to its formation around the British Marconi radio interests and patents). Once the officers of the BBC saw the sort of sharp pictures that electronic television could deliver, they expressed their dissatisfaction with Baird's system and invited him to conclude his experiments. This development came as a surprise to Baird's backers, who urged him to develop an electronic television system of his own so that he could stay competitive. Baird was steadfast in his resistance to the suggestion, so the directors of British Gaumont, ignoring the objections of the inventor in whose name they acted, sent a team of engineers to America to visit Philo T. Farnsworth.

By the time Farnsworth set sail for Southampton, the BBC was ready to abandon mechanical scanning altogether and convert its experimental television operations over to an electronic

John Logie Baird with the prototype of his "Televisor"

television format based on the system that EMI had introduced. The receiving end of the EMI system was a familiar cathode ray tube, similar to Farnsworth's Oscillite and Zworykin's kinescope. The camera tube, which EMI immodestly dubbed the "Emitron," was a much more intriguing apparatus.

What is curious about the Emitron tube is its unmistakable resemblance to another device, the Iconoscope, which Vladimir Zworykin had first demonstrated for RCA in the early 1930s. Both tubes employed the same lopsided geometry, built around a single-surface photo cathode composed of discrete photoelectric islands, and an oblique, triangular scanning configuration. There is no question that the Emitron and the Iconoscope were the same device. The only question that remains unanswered—to this day—is which laboratory really produced it first?[70]

In other words, as Farnsworth sailed for Europe hoping to form an alliance that would enable him to overcome his difficulties at home, his principal domestic adversaries were already operating a trans-Atlantic alliance of their own. Of course, Farnsworth knew nothing of all these backstage machinations. As his boat arrived in Southampton, he was unaware that the struggle to bring television to Europe would be drawn along exactly the same lines as his struggle in America.

Aboard the S.S. *Bremen* with Farnsworth and his precious cargo were two laboratory assistants, Tobe Rutherford and Arch Brolly, as well as Skee Turner, whose role was to assist Farnsworth in the negotiations for a patent license. Unfortunately, the *Bremen* was a German ship arriving in Britain at a time when relations between the two countries were already tense, to the point that German ships could not be unloaded in British harbors. Consequently, all the cargo had to be transferred from the ship to a smaller British vessel before it could be unloaded onto the dock.

While Farnsworth and Turner went off in a launch to meet their hosts from Baird, Tobe and Arch stayed on board to keep an eye on the equipment as it was lifted by crane from the cargo hold. This turned into an unexpectedly tricky maneuver. Gusty winds and choppy water caused the crate holding Farnsworth's precious equipment to sway precariously as it was hoisted out of the hull of the *Bremen*.

Tobe and Arch watched anxiously as the crane swung out over the edge of the big ship and lowered the crate toward the bobbing deck of the smaller transfer ship. The crate was only inches from a safe landing when a sudden wave caught the smaller vessel; instead of the crate being lowered gently to its deck, the deck rushed up to meet the crate, smacking it with a force equivalent to a fall from several feet.

Not until several hours later, when the crate was unsealed at the Baird headquarters in London, were Farnsworth and his men able to assess the damage. John Logie Baird stayed alone in his office, but his representatives hovered about restlessly while Phil and Tobe lifted the lid. The sudden change in their expressions when they peered into the crate was a dead giveaway that things inside looked grim. Indeed, three racks of electronics had sheered away from their mountings and fallen into a heap at the bottom of the crate. Baird's men smiled to each other when they saw the mess for themselves—now they could report to their boss that the American machine was wrecked.

Less than an hour later, whatever slim cause for celebration Baird might have had was interrupted when Phil appeared at the door and invited Baird and his men back downstairs for their first look at electronic television. The Baird contingent followed Turner downstairs, joined along the way by executives from British Gaumont.

As he had done at the entrance to the Franklin Institute exhibit, Farnsworth had placed both the camera and the receiver near the entryway. The instant Baird and his associates entered the room they were confronted by their own disembodied image, rendered in stunning clarity and detail.

The Britons were startled by the experience. After years of financing Baird's mechanical television system, the most resolution that the British Gaumont people ever saw was 60 lines per frame. Now they were confronted with an image composed of more than 300 lines per frame, delivering detail their man had always assured them was quite impossible. Confronted by Farnsworth's obvious accomplishment to the contrary, the British Gaumont people realized that they'd bet on the wrong horse.

The stunning effect of this demonstration did little to prepare the hosts for the shock of Farnsworth's terms. The board of directors of British Gaumont sat in bemused tolerance while Farnsworth and Skee Turner explained that in addition to the customary continuing royalties, he wanted a $50,000 down payment to accompany the license, as a sort of opening fee, a payment of royalties-in-advance.

What British Gaumont had envisioned was more like a mutual exchange, a sort of our-patents-for-yours proposition, with no cash involved. But Farnsworth couldn't think of anything in the British Gaumont patent portfolio that he really needed, certainly none of John Logie Baird's mechanical television patents. It seemed to Farnsworth that he was the only one holding any cards in this game, and he stood firm: $50,000 cash or no license. The negotiations bogged down.

Seeing the impasse, Farnsworth addressed the representatives for British Gaumont. "Gentlemen, we don't seem to be making much progress here." Gesturing toward Skee Turner, Farnsworth said, "Will you excuse us for a moment, I believe we need a short recess so I can confer privately with my associate."

Farnsworth and Turner excused themselves to an adjacent room. Once alone, they hardly needed to speak; the determined look in their eyes was mutual. They were not going home empty-handed.

Skee Turner was the first to notice a bottle of Scotch and one small glass standing on a mantel, and with a compulsive, defiant flourish, he poured himself a shot and choked it down. Skee then handed the bottle and glass to Phil, who did the same, hesitantly. After a brief moment to recompose themselves, Farnsworth and Turner returned to make their final stand before the British Gaumont Board of Directors.

It's not hard to envision the subsequent encounter: the spokesman for British Gaumont leans forward, confident that these young, inexperienced bargainers were about to propose a clever Yankee "compromise." They were about to discover instead just how indomitable these two Yankees could be after they'd been fortified with a shot of single-malt Scotch.

"We want cash," Farnsworth declared, pounding his fist on the table in a gesture that surprised everybody, including Farnsworth. The tone of finality in his voice assured the Gaumont officers that the negotiations were about to conclude, one way or another. The Gaumont men stiffened in surprise, mumbled among themselves with proper British indignation for a few moments, and then conceded to Farnsworth's demand.

As exciting as their cruise across the Atlantic to Britain must have been, the return voyage with $50,000 in their pocket must have been truly exhilarating for Farnsworth and Company. After all, that sum represented their first genuine reward for nearly ten years of concentrated effort. At last, they had an honest-togoodness, royalty-paying patent license in their possession. Others were sure to follow.

Motivated by their sudden achievement, Farnsworth and Skee Turner spent the entire week at sea daydreaming about ways to parlay their windfall into even greater success.

The first public demonstration of Farnsworth's system a few months earlier at Philadelphia's Franklin Institute, where thousands had lined up to see the new electronic marvel, proved that television was ready for the public and that the public was more than ready for television. The next step, Phil and Skee decided, was to demonstrate the day-to-day feasibility of television broadcasting—something which they could do immediately by investing the British license fee into building a fully equipped television studio that could sustain a regular schedule of experimental broadcasts.

Of course, the $50,000 license fee was not entirely Farnsworth's property. The money, as well as all of Farnsworth's patents, were actually the property of Farnsworth Television, Inc., the new holding company that Jess McCargar formed to raise money when Farnsworth walked out of Philco one year earlier. Although Farnsworth still owned a significant portion of the equity in the enterprise, he was by no means the majority stockholder. Instead, the numerous investors who had acquired chunks of stock were represented by their own board of directors, and that august body, not Farnsworth, would determine the disposition of the Baird license fee.

From almost the moment he stepped off the gangplank, Farnsworth realized how much the power in his life had shifted. Jess McCargar demanded that the $50,000 be forwarded immediately to the business offices in San Francisco, and that the matter of building a studio be tabled until the board of directors could consider it. Jess figured that $50,000 would serve nicely in the meantime to pay off some old bills, such as the $30,000 tab for legal services contracted during the patent litigation with RCA.

Farnsworth and Turner saw their dream of a studio facility— one that would give them an entry into the lucrative broadcasting business—crumble in the face of McCargar's shortsighted single-mindedness. What Farnsworth feared more than anything was that his backers would take a conventional approach, attempting to secure the future of Farnsworth Television by doing what Phil called "tacking on the shipping room door"—that is, following in the pattern established by the giants like RCA and Philco, opening their own factory to build and sell merchandise like television and radio receiving sets.

Phil could certainly see the expedience in such an approach, as well as the inherent risks. But it was a far cry from the purely creative "invent and license" enterprise that he had been dreaming of since he'd first confided to his father his hope that he had been "born an inventor." It was beyond frustrating that, even with a royalty advance of $50,000 in hand, McCargar still didn't get it.

If a more conventional enterprise had to be formed around the Farnsworth patents, then Phil and Skee Turner were certain that a far more lucrative future could be found, not in manufacturing, but in the field of broadcasting, based on the model that commercially sponsored radio broadcasting had already proven to be fabulously profitable. Building a studio facility seemed like a logical step in this direction and a viable alternative to manufacturing. In his dreams, Farnsworth would let others worry about operating factories while his company just collected patent royalties for the products that others manufactured.

However, the debate over future business models never really surfaced when confrontations began to erupt with Jess McCargar. For his own part, Jess was merely questioning the wisdom of taking on another sizable expense when they could hardly meet current expenses. But when the board of directors sided with McCargar, Farnsworth realized that the odds were no longer in his favor. As much as he hated to admit it, this denial of a studio left Farnsworth numb with the realization that—not unlike John Logie Baird—he had lost control of his destiny to men who could not share his vision.

ELEVEN

Something for Nothing

The spherical Multipactor tube, ca. 1935

"Sometimes one pays most for the things one
gets for nothing."
—Albert Einstein

Farnsworth and Skee Turner decided to go ahead with their dream of building a television studio that would be physically and financially separate from the Mermaid Lane laboratory, regardless of the response of Farnsworth's other backers. Turner felt strongly that such a facility was essential if Farnsworth was ever to surmount the day-to-day problems of commercial television. He was so certain that commercial broadcasting offered the prospect of a lucrative future that Turner persuaded his father to put up the money.

Turner also designed the studio, which was located in a prefabricated structure on a hill in the Wyndmoor section of Philadelphia.

While contractors outfitted the empty building with a stage and lights, Farnsworth and the lab gang devoted their time to constructing two state-of-the-art Image Dissector cameras. These cameras were built for durability as well as for high resolution. All the equipment was engineered so that both the camera and the picture tube would be capable of producing 441 lines per frame—well in excess of the 400 lines per frame that Farnsworth had established as his objective back in San Francisco in the late 1920s.

Farnsworth's crew created and built a special transmitter and a 100-foot tower that could blanket the Philadelphia metropolitan area with experimental television signals. They also designed and built the world's first electronic video switcher, which allowed instantaneous intercutting between the two cameras as programs were broadcast. This technique allowed a television program director to replicate in real time the editing technique that film directors use to cut between wide angles or close-up shots. While all the equipment was under construction, the FCC granted Farnsworth a license to conduct experimental television transmissions, and as soon as construction was complete, W3XPF was "on the air."

One big job for the new broadcasters was arranging with artists to perform before the electronic eye. Farnsworth had learned during the Franklin Institute public demonstration of 1934 that he could rely on an endless supply of amateur singers, dancers, musicians, and magicians willing to trade their time for a little exposure and the undeniable thrill of being televised.

What programs they were able to assemble were broadcast to a handful of prototype receivers that were beginning to appear around Philadelphia. There were still no television sets for sale to the general public, but there were three companies in the area experimenting with electronic television: Farnsworth, RCA, and Philco. Many of the engineers who worked for these companies had experimental receivers in their homes that they used to monitor transmissions from their own laboratories. These few dozen homes that were actually equipped with TV became focal points of the neighborhood, truly the first on their block.

The studios of W3XPF, near Philadelphia, in 1936

Television was getting started, for the time being at least, in much the same way as radio got its start—in the hands of a small legion of clever and skilled enthusiasts who could build their own receivers to catch whatever signals were dancing in the ether. But this informal, pioneer spirit would not last for long. Television, which is much more technically complex than sound-only radio, has many more variables in its scan rates, frame rates, broadcast bandwidth, and other technical aspects. Consequently, television would require that its industrial advocates arrive at a single set of standards, so that all TV sets could receive all the programs broadcast within any given area. It would never do for RCA to broadcast with one set of signal standards while competitors like Farnsworth or Philco used another. The prospect of pulling the competing interests of the industry together to negotiate such standards was one more obstacle that Farnsworth would have to contend with before he could unleash the value still tied up in his patents.

Some idea of just how protracted the commercial realization of television would be was revealed in a *Newsweek* article that appeared in August 1935. The difference between Farnsworth's outlook and that of his principal adversary was spelled out in the headline: "Television: 10 Years to Go, Says RCA; 1, Insists Inventor." In the article, David Sarnoff was quoted, declaring that a nationwide system of "home television" might require ten years to develop. "But last week," the article continued, "into the television picture popped Philo T. Farnsworth with a visionary gleam in his blue eyes." The article then quoted Farnsworth directly: "Technically, we are ready to begin home production. It won't begin until the receiving sets can be sold for $200 or less. I expect that can be accomplished within a year."[71] Clearly, more than just patent interferences were separating the visions of David Sarnoff and Philo T. Farnsworth.

Phil demonstrates a prototype home receiver for his secretary, Mabel Berstein

Meanwhile, the Wyndmoor studio, as was intended, revealed just what sort of difficulties the industry might encounter should commercial broadcasting begin anytime soon. Crews in the studio discovered that the current versions of the Image Dissector tube had some unexpected infrared sensitivity, which caused the color red, which normally photographs black, to televise as white. This was not unlike the problems that surfaced with early motion-picture films. So the Max Factor Company in Hollywood—looking to secure their own beachhead in the new art of television—was happy to lend its expertise. The coloration problems were solved by applying blue makeup around the performers' lips and eyes. The solution worked great on television, but the performers often looked ghoulishly blue to the eye.

The Image Dissector also displayed a peculiar sensitivity to certain fabrics, rendering them transparent, as was graphically illustrated the day a pretty ballerina seemed to be dancing naked on the video screen.

Despite these inconveniences, video broadcasts from the Farnsworth studios not only proved the feasibility of television but also gave Farnsworth a public image. Though fewer than fifty

The Max Factor Company provided makeup
solutions for the Farnsworth cameras

homes in the Philadelphia area were equipped with video receivers, that activity was enough to add the word "television" to the common vernacular. And the boy genius from Utah became a minor celebrity.

For a few months, Farnsworth enjoyed the attention; it was a welcome change after nearly ten years of hard work in total obscurity. But eventually the obligations of even minor celebrity status became too pressing a burden on his time. What with meetings and interviews and visits from foreign dignitaries all hoping to spend a few minutes with the great Farnsworth, being famous became a job in itself, entirely apart from the work that made him famous in the first place.

Besides, Farnsworth had ideas of his own that needed pursuing. By the winter of 1935, he had worked out the particulars of his "super amplifier" tube, and once the patent[72] was filed, he began building the device itself.

Again demonstrating his penchant for obvious nomenclature, Farnsworth called the new device the "Multipactor" because it drew its power from the multiple impacts of secondary electron emissions. George Everson described the Multipactor as "a cylindrical tube sealed at each end with two disks of metal coated with cesium opposing each other at equal distances from the ends."[73] The tube was suspended in an oscillating magnetic field, and current was applied to the cathode plates. Unlike most electron tubes of the time, which released electrons when their elements were heated, the cathodes in the Multipactor were operated cold.

The results were astonishing. Because the tube operated "cold," there was no evidence that it was actually working, yet it produced "amazing amplification of power, apparently coming from nowhere." In an uncharacteristically effusive letter to Donald Lippincott, Farnsworth said the Multipactor "was like getting something for nothing."[74]

The significance of the Multipactor cannot be overstated. Though far less glamorous than television, the Multipactor was in its own way an equally significant breakthrough. Because its effects are not as obvious—nor as widely anticipated—as television, the Multipactor has often been overlooked in the annals of technology. But it was truly a revolutionary device. Reporting about the Multipactor in the *San Francisco Chronicle*, science writer Earle Ennis wrote:

> ...an astonishing new radio-television tube that not only transmits television impulses, but may be used as an amplifier, detector, rectifier and multiplier tube as well, and may make obsolete all known forms of radio tubes, was announced yesterday by the Television Laboratories, Inc. The tube is the long sought "cold cathode" tube which has been the goal of laboratories the world over. It is a multiplier of current to an astonishing degree and because of its five-fold function it is of worldwide scientific interest.

> To make plain the operation of this child Titan without going into a technical description, it must be understood that modern radio tubes are all of the "hot cathode" type. The source of the electrons is a filament. When this filament is heated by passing a current through it, the electrons are "boiled off" by high

temperature. The process is scientifically ineffective, as a comparatively large amount of power is required to obtain a small number of electrons. The "cold cathode" tube, long sought as a solution to radio troubles, is one that has no filament or grid, and so has nothing to burn out. It has no instability due to changes in gas pressure and is, in a sense, perpetual and indestructible. In a "cold cathode" tube there are two plates between which the electrons pass. In the Farnsworth tube they are "bounced" with terrific force on the two plates. In this "bouncing" lies the secret of the tube's amazing new powers.

The "bouncing" process knocks additional electrons from the cathodes, and these in turn are bounced against the electrodes, which in turn have more electrons knocked out of them. In the Farnsworth tube, a single electron will build up or father 2×10^{60} [that's the number 2 followed by *sixty* zeroes] electrons, all in the space of 1/1,000,000th [one one-millionth] of a second. Each of the "children" is a perfect electron, with all the properties and speed of the parent. This, it can easily be seen, produces a tremendous increase of current, so great in fact that if it is not drawn from the tube inside of the 1/1,000,000th of a second, the tube electrodes melt.[75]

Beyond such abstractions, the power was demonstrated when the Multipactor was used to relay radio messages around the world, from San Francisco to Australia, through several stations in Europe, and finally reaching Phil and Pem in Philadelphia.

Besides its usefulness in radio applications, the amplification provided by the Multipactor tube proved a tremendous boost to the Image Dissector, making it much more practical for studio and outdoor scenes than earlier versions. But the Multipactor would, over time, prove to be a far more important discovery than the mere improvement of a television picture suggests. Moreover, the Multipactor was precisely the kind of unexpected breakthrough that Farnsworth always knew would emerge from his laboratory as long as he was given the means to follow his imagination and let ideas gestate in their own good time, free from the interference of business necessities. The Multipactor evolved from an idea that grew out of an observation that Farnsworth had made early in his work. The need for amplification of television signals stimulated Farnsworth's interest in secondary emissions, which eventually produced the Multipactor tube. What else might the fertile atmosphere of Farnsworth's laboratory lead to? Only time and a stable, adequately funded laboratory would tell.

The Multipactor went through numerous configurations during the thirties. One is of special interest. This particular variation featured a spherical formation, rather than the typical cylindrical tube. Two hemispherical electrodes were placed near the perimeter of the glass sphere, and another electrode—a wire grid—was placed in the center. When Farnsworth operated this spherical Multipactor at relatively high levels of power, he observed a tiny, brilliant, blue, star-like anomaly suspended within the inner grid in the center of the tube. As the power levels in the tube were increased, the point of light became more brilliant. Even more impressively, it never touched the walls of the tube itself. The phenomenon seemed to suspend itself within the glass walls of the tube.[76]

Farnsworth was not sure what to make of this star-like phenomenon when he first observed it, but he suspected that it was important. Just as he had recorded his earliest observations of secondary emissions and stored them in the back of his mind for future reference, he made particular note of the blue star. Instinctively, he knew he would be returning to it some day.

The Multipactor, just one of more than a dozen new patents that Farnsworth filed in 1935, represented a stunning breakthrough among the technical cognoscenti. Still, it was television that stimulated the public's interest. The flood of publicity coming out of the Wyndmoor studio peaked in 1936 when the Paramount Newsreel Service, the "Eyes and Ears of the World," ran two stories about the coming of television and the remarkable man who put it all together.[77]

By raising the visibility of the Farnsworth company, nationwide publicity raised the value of the stock, and the value of everybody's holdings swelled appreciably. Some accountant with a sardonic wit told Farnsworth that at current prices, his own holdings were worth more than a million dollars. These figures made Farnsworth a living example of the American dream—a millionaire before his thirtieth birthday. Of course, this was only a "paper" fortune. The fact was, Jess McCargar and George Everson continued having difficulty finding investors willing to buy into a company that still could not sell its only product.

In October 1936, *Colliers* ran a feature story about "Phil the Inventor" that echoed the sentiments Farnsworth had expressed in the earlier *Newsweek* article, saying that television seemed destined to find its way into many American homes by Christmas 1937.[78] This prediction reflected a common feeling of the time that commercial television was "just around the corner." But Farnsworth knew that as long as his patents remained under contention, as long as the debate over signal standards was stalled, and as long as events unfolded at the pace determined by existing industrial interests, turning corners would take much longer than any magazine writer could predict.

Nonetheless, with the job of perfecting and promoting electronic television off to a strong start under his own roof, life at Farnsworth's Mermaid Lane laboratory took on a new dimension. Many of the men still working for him had joined him years earlier in San Francisco. Under his youthful guidance, this unlikely group managed to turn the tangle of wire and glass that produced the first electronic television picture into what the Paramount newsreel called the realization of "mankind's most fanciful dream." For Farnsworth, the lab itself was the realization of his own most fanciful dream.

Others before them had failed, crying that it could not be done without massive infusions of capital, but Farnsworth and his lab gang proved them all wrong. They not only invented TV, they overcame the limitations of their financing and delivered their invention to the marketplace, ready for the start of commercial broadcasting. After so many successful years together, the lab gang began to take on the air of scientific invincibility. The unwritten motto for the entire operation was "The difficult we do right away. The impossible takes slightly longer."

Among the dozens of new patents that Donald Lippincott filed during 1935 and 1936, Farnsworth was most proud of those attributed to other members of the lab gang. These filings reflected the collective spirit that Farnsworth instilled in his co-workers. Farnsworth's generous, straightforward approach to the work provided an incentive that tied the lab gang together. Their hours were long, the work was sometimes tedious and painstaking, and the pay was never abundant, but Farnsworth never had any trouble finding capable men who were not only willing but also eager to work with him.

As the lab gang grew, Farnsworth chose new men very carefully, watching closely for people who displayed both compatibility and trainability. Admission to the lab gang was predicated primarily on an applicant's willingness to take chances. What kept the lab going was men who could, by following Farnsworth's example, find their own way of doing whatever they'd been assigned and make it work. In this manner, Farnsworth built a well-organized team that could deliver the specific ingredients of his designs. This was the vortex of creative and scientific energy that would deliver not only television to the world, but whatever else Farnsworth set his sights on creating.

Once accepted into the gang, a new employee found himself welcomed into what Tobe Rutherford called, "one big happy family." Indeed, many lab workers were members of Farnsworth's immediate family: his brothers, Carl and Lincoln, were both members of the lab gang. His sister, Laura, studied with Max Factor and became the resident makeup consultant. And his chief tube-builder was his brother-in-law. This extended family became a collective unit that was the extension of Philo's incredibly creative scientific mind.

At the same time that he was directing members of his research team to solve particular problems growing out of the day-
to-day television operations, Farnsworth concentrated his own
attention on the questions of basic science which his instincts led
him to explore. The solid support of his laboratory group supplied the platform from which he could launch these explorations
toward the outer edges of electronics and physics where new discoveries might be revealed.

Over the years, Farnsworth developed a unique methodology
both in and out of the lab. Usually, in an environment like the
Farnsworth laboratory, it is very easy for workers to get side-
tracked, to steer their time and energy down blind alleys. That
rarely occurred at the Farnsworth lab. Phil had an uncanny ability
to see a destination and guide his men there. Where others might
be led astray, Phil could make side trips without breaking his
stride; these diversions supplied the observations and discoveries
that shaped his thinking. "He could go through the orchard and
just pick things out without breaking his rhythm," one observer
noted. The only thing that ever slowed him down was the short-
age of funds.

Cliff Gardner marveled once that "the amazing thing about
Phil was his ability to invent solutions without even realizing he
was inventing. He was just providing solutions to the day-to-day
obstacles we'd encounter. He always had an answer when one of
us asked 'what do we do now?'"

Over the years, Farnsworth trained his mind, fine-tuning his
innate intuition. Often, he would bring a problem home with
him to mull over. Pem would see him, as the Sun went down,
sitting at the piano or playing his violin in the solarium of their
home. Sometimes she would see him playing solitaire. He called
such activities a "counter-irritant" that would focus the front of
his mind while the back burners did their magic. Sometimes he
would just concentrate on a problem before he went to bed,
wake up—often in the middle of the night—with a perfect solution. "He'd show up at the lab the next day," Pem remembered,
"and he'd pull one of the workers aside and say do this or that

or the other, and when it worked they just couldn't get over it, it just seemed like magic to them. They never realized all the extra work he did."

The Multipactor tube was only one case in point. With nearly ten years of firsthand experience on the invisible frontier, Farnsworth became convinced that there was no limit to the things he could get electrons to do. He turned to his lab gang to construct the tubes and circuits that could prove his point.

Electronic television, which began as a dream in the mind of a teenager, had become a virtual reality, but the lab gang was the fulfillment of an even grander dream. Television was as much the product of their sweat as it was the gift of his genius. Farnsworth was the dreamer; the lab gang was the instrument of his dreams. With these men at his side, whatever Philo T. Farnsworth wished of the future was at his command.

TWELVE

You're All Fired

The Crystal Palace in Kensington Gardens, London

"I never think of the future. It comes soon enough."
—Albert Einstein

Throughout 1935 and 1936, Farnsworth carried a staggering workload. Typically, he spent his mornings at the Wyndmoor studio, personally conducting demonstrations for the daily tide of visiting dignitaries who felt they were somehow entitled to spend a few minutes with the now-famous Philo T. Farnsworth. Afternoons he spent at the laboratory on Mermaid Lane, working on solutions to problems that came up at the studio and elsewhere in the relentless pursuit of commercialized television. Evenings he spent either at the lab or at home, working on new ideas and developing the mathematics for his own projects. Despite the never-ending demands on his time and attention, this period was incredibly productive; Farnsworth and his team filed more than two dozen patents for the Multipactor and other innovations that had been kept under wraps during his final months at Philco.

These jobs alone were enough for three men, but there was no end to the additional distractions that kept Farnsworth away from what he considered his most important work. Equally distressing, the people who were primarily responsible for funding Farnsworth's enterprise were suffering their own strains, trying to raise enough money to maintain the operations and cover a payroll that—between the lab and the new studio—provided livelihoods for more than two dozen employees.

Given the immediate need of keeping the venture together until television could be unleashed on the marketplace, it is not surprising that, back at corporate headquarters in San Francisco, Jess McCargar and the stockholders he represented did not share Farnsworth's enthusiasm for opening new lines of research. There was still no settlement in sight from RCA, whose engineers were now using Farnsworth's inventions to improve their own system of television with little regard for such pleasantries as patent licenses or royalty payments. Until the RCA situation was cleared up, all this talk of new ideas struck Jess McCargar as premature.

In fact, McCargar expressed considerable impatience with the whole affair, and suggested on more than one occasion that maybe accepting RCA's offer for a complete buyout wasn't such a bad idea—which, of course, is precisely what David Sarnoff was holding out for. Jess's increasingly frequent suggestions that Farnsworth take whatever deal he could from RCA or AT&T or any other willing buyer only proved to Phil that Jess would never understand how a patent could work like an annuity, paying dividends for as long as the patent was in effect without ever being sold. McCargar should have been able to grasp this concept. He was, after all, a banker with years of experience in financial matters. Here he was, holding the reins of a whole new industry, but McCargar was hardly an industrialist in the manner of George Westinghouse or Henry Ford, or even David Sarnoff. At best, Jess McCargar was a deal-maker. At worst, he was simply a stockbroker who could make judgments based only on how much things cost. Where the lab operations were concerned, they always cost too much.

In the closing months of 1936, all these pressures began to have a noticeable effect on Farnsworth's increasingly delicate physiological balance. He began to come home tired each day, and his disposition, usually cheerful and optimistic, sometimes turned sour. Pem began to take notice of an alarming new habit: At the end of each day, Phil was gulping down a highball or two before dinner, often followed by one or more during the evening so he could settle his nerves enough to get some sleep. After pushing himself relentlessly for ten years, Phil Farnsworth was beginning to reach the limits of his endurance—and was starting to medicate himself to relieve the strain.

As if life for the Farnsworths wasn't already intense enough, Phil learned in the autumn of 1936 that the Baird operation in England—still his only licensee—was in trouble. With his continued reluctance to adopt electronic television, John Logie Baird had made such a mess of things that the BBC was all set to award its television contract to EMI. But Baird's corporate sponsors raised such a fuss that the matter came before Parliament, where the Selsdun Committee was appointed to make certain that the BBC conducted competitive testing between EMI and Baird before awarding the contract. As the tests got underway in 1936, Baird was having a bit of a problem with his Image Dissector tubes. Baird, who had done more to advance the dinosaur art of mechanical television than any other experimenter, couldn't get a picture with his new-fangled electronic camera tubes.

Unfortunately for Farnsworth, there was no one else in the world Baird could turn to for help. Aside from being tired and stressed, the American inventor was understandably reluctant to leave the lab lest McCargar do something precipitous during his absence. But since his only industrial ally was on the verge of collapse, he had no choice. He agreed to return to Europe, but only on two conditions: he insisted on taking a slow boat so that he could have a few days to rest; and he insisted that additional passage be provided for Pem. Baird accepted these conditions, and Mr. and Mrs. Farnsworth sailed for Europe, making a honeymoon of it ten years after their wedding.

In London, Farnsworth found
Baird's equipment set up in the elegant Crystal Palace in Kensington
Gardens, the remarkable glass and
steel edifice that Queen Victoria had
built in the 19th century to celebrate
Britain's role in the Industrial Revolution. Farnsworth was shocked to
discover the source of Baird's problems: two years after taking a license from Farnsworth, Baird was
still using a scanning disk for certain components of his system. True,

John Logie Baird

Baird had pushed his mechanical creation to the point where it
could deliver some 200 lines per frame, but the competing system
offered by EMI produced 405 lines per frame, and clearly left
Baird standing in the dust.

Baird was using his Image Dissector tube for his "cinefilm"
transmitter, which is the British euphemism for a film-chain, but
even in this capacity he was not taking full advantage of the Dissector tube's capacities. In fact, Farnsworth found the chassis for
the Image Dissector only partially built; Baird and his men simply
didn't know how to finish it.

Once he was on the scene at the Crystal Palace, Farnsworth
plunged into an around-the-clock schedule to put Baird back in
business. When he was done, Phil and Pem drove with Baird and
members of the Selsdun Committee to a small pub outside London, where a cathode ray tube receiver was set up to catch
Baird's over-the-air transmission. Everybody saw the crisp detail
and subtle contrasts in the picture delivered by the Farnsworth
Image Dissector.

Still, there was little Phil Farnsworth could do to help John
Logie Baird as long as the Scotsman hung on to his spinning
wheels. Evidently, British Gaumont, Baird's backers, felt pretty
much the same way as Farnsworth. It seemed that Baird's
chances of winning the BBC contract were doomed even before

Farnsworth arrived at the Crystal Palace. Nevertheless, Phil worked tirelessly to get Baird back up to speed, but Pem could see that the effort, once again, was draining him of his reserves. Phil and Pem started to think about where they could go to get some sun. The French Riviera sounded nice.

Then a wire from Cliff Gardner confirmed Phil's worst fears about leaving the lab unsupervised. In his absence, Jess McCargar had dispatched an associate of his, another former stockbroker named Russ Pond, to Philadelphia to take over the day-to-day management of the lab. Pond's qualifications for the job hardly fit the requirements. He was not an engineer, not an accountant, not even an office manager. He was a securities dealer, what in the trade was often referred to as a "customer's man." As such, he was a perfect stand-in for Jess McCargar, and his presence proved an irritant to the workers.

Phil's first impulse upon hearing this news from Cliff was to drop everything in Europe and return immediately. Phil could tell in his bones that a confrontation with McCargar was looming, but Pem persuaded him that he was too run-down to risk such a showdown before he got his personal batteries recharged. The Riviera was sounding better all the time.

Phil wired Cliff back, asking him to take a firm stand where McCargar and his surrogates were concerned, then he and Pem boarded the elegant "Blue Train" heading for the south of France.[79]

Once on board, the Farnsworths discovered they were traveling at a unique time in European history. All the gossip in the dining car revolved around the recently exposed intrigue of Edward VIII's illicit romance with the American divorcée Wallace Simpson. The Farnsworths arrived on the Riviera at almost the same time as Mrs. Simpson, who had been forced to flee London as word leaked out of her relationship with the King.

Their destination was the village of Mentone, near Nice, and the Hotel de Orient overlooking the Mediterranean Sea. Pem was so concerned for Phil's health that she asked a local physician to examine him. With a friend acting as interpreter, the doctor prescribed some medication to help Phil relax—and at least two weeks of rest. Actually, the doctor wanted him to spend those two weeks resting in bed, but Phil wasn't about to spend two weeks on the coast of France looking out a window. Instead, he hired a chauffeur and spent three weeks leisurely touring the countryside and exploring the towns of Nice, Monte Carlo, and San Remo.

Those three quiet weeks provided a much-needed respite for both Phil and Pem, their first real vacation after more than ten years of hard work, and their first time off together after ten years of marriage. As Phil's health improved, the backdrop of the French Riviera provided just the right romantic touch to reawaken the fire of their affection. He gained strength steadily, but as Pem pointed out, "the strain of the past ten years could not be erased in a week or two."[80]

While they were on the Riviera, the Farnsworths were shocked to read one morning that the glorious Crystal Palace had been leveled by a fire of mysterious origins, which destroyed all of Baird's equipment, including Farnsworth's Image Dissectors. The fire brought almost total ruin to John Logie Baird, and all but guaranteed that EMI would win the contract to put the BBC into electronic television.

The news was a setback to the positive energy that this trip seemed to be gathering. Phil and Pem returned to London to survey the damage. Sifting through the rubble of the Crystal Palace, Phil found a macabre souvenir of the tragedy—the charred, melted remains of his Image Dissector tube, which he carefully placed in one of his bags, to be carried home with him as a grim reminder of what was left of the British hope.

After a few more days in England, Phil and Pem flew to Berlin to accept a long-standing invitation to visit with Dr. Paul Goerz, president of the German electronics company Fernseh, AG. Dr. Goerz was one of the many foreign dignitaries who had visited Farnsworth's lab at Mermaid Lane; Farnsworth and Goerz had become friends, and Fernseh had access to Farnsworth's patents through a cross-license with Baird, but Phil went to Berlin to find out why Fernseh was not paying the royalties required under that agreement. Once the matter of royalties was settled, Fernseh used Farnsworth's system to televise the 1936 Olympic Games—the infamous "Nazi Olympics"—later that year in Berlin.

Farnsworth sits for a formal portrait on his 30th birthday

The entire trip to Germany was conducted under a cloud of intrigue and foreboding. The Nazi regime was everywhere in evidence, and Phil and Pem got the distinct sense that Der Führer would like them to remain in Germany indefinitely. They were assigned military escorts for drivers, who, they learned later, were reporting directly to the S.S., the secret police. And when it came time to leave, Dr. Goerz had to make special arrangements, practically smuggling his guests out of the country.

Following their narrow escape from the Nazis, Phil reluctantly came to the conclusion that a major military conflict in Europe was inevitable. He expressed to Pem his sincere, heartfelt hope: that television might be used somehow to minimize the consequences of any conflict, by bridging the gap between nations who might soon find themselves at war. For that to happen, television would first have to find its way into the marketplace. It was hard to tell from the vantage point of the mid-thirties which would happen first.

After a stormy, gut-wrenching passage back to the United States aboard the S.S. *Europa*, six weeks after running to Baird's rescue, Farnsworth looked forward to his return to the lab with some trepidation. Not only was he concerned about the correspondence with Cliff and what he might find morale-wise among the lab gang, but he was also worried about the fate of some patents he had left to be filed prior to his departure.

He was particularly concerned about a patent he had filed for a process known as "velocity modulation." Simply stated, velocity modulation is about manipulating the flow of electrons in a vacuum tube so that they arrive at their target in bunches. Hidden within that over-simplification is the revelation that velocity modulation is one of the rudiments of a then entirely new science called "Radio Detection and Ranging"—or RADAR for short. Especially after what he had seen in Germany, Phil suspected that radar was going to be a very important field. He was equally certain that the patents he had filed could be as fundamental to the development of radar as his 1930 patents were proving to be to the development of electronic television.

Again, velocity modulation was precisely the sort of advanced work that Phil knew would have value in the future, but only if the proper claims were filed at the time of their discovery. Such breakthroughs grew logically out of the work he and his team were doing in television. Even with competition in the television arena nipping at their heels, the Farnsworth lab was far enough ahead of the rest of the pack that new fields like this just naturally opened up to them first.

But patents had to be filed and the filings had to be supported with supplemental materials and all that took attorney time and that cost money. New art like velocity modulation represented precisely the kind of "advanced research" unrelated to television that required the sort of added expense that Jess McCargar was no longer willing to bear.

Though not a man ordinarily prone to emotional extremes, Farnsworth found it difficult to contain his disappointment when he learned that McCargar had allowed his velocity modulation patents to lapse while he was in Europe. Whatever claim Farnsworth might have had in the emerging field of radar was lost as others rushed in to fill the void—and file valid patents.

Things were no better with the rest of the operation. Phil learned that Russ Pond, Jess McCargar's stand-in at the lab, had taken his new job very seriously, but showed little regard for the intricacies of science or the delicacies of an operation like the Farnsworth lab. The result of his arbitrary management style was a badly demoralized lab gang. After nearly ten years together in the trenches, everybody was grumbling about how things had fallen apart in the six short weeks while Farnsworth was in Europe. The entire lab gang was suffering from a case of badly damaged esprit de corps.

Anticipating the coming confrontation with McCargar, Farnsworth took matters into his own hands. He advised Russ Pond that his services were no longer required and sent him packing. He then called Jess and told him what he had done with Russ Pond. Four days later McCargar himself showed up in Philadelphia. The collision Phil had been dreading was about to occur.

All the members of the lab gang saw McCargar arrive. They put down their tools and glared at him as he stood in the doorway. They practically saw the smoke pouring out of his ears as he nodded around the room, conducting a silent head count. Then McCargar stormed into Phil's office.

"Phil," Jess steamed, "this operation is just too top-heavy. You've got more men than you need out there. You've got to let some of them go."

"No, Jess," Phil answered calmly, "you're wrong about that. This is a carefully assembled team, and each of those men plays a vital role in our operation. I'm not firing anybody."

"Well then," Jess shot back, suggesting a familiar tactic, "if you can't fire some of them, fire all of them and hire back just the ones you really need."

Farnsworth had been down this path before. This, he decided, was the place to plant the flag. "No, Jess," Phil calmly replied. "I won't do that."

"Either you go out there and fire them," Jess fumed, "or I will!"

"Calm down," Phil implored, "and think about what you're doing. Each of those men is essential to some aspect of our work. If you let them go now, we'll lose ground to the competition, and everything you've invested will be lost. I'm not going to fire anybody."

"If you won't do it, I will." McCargar fumed and turned his back on Phil. He walked back into the lab, gesturing wildly and yelling, "You're all fired!"

The men all looked at each other in stunned silence as McCargar raged on.

"You heard me. Get your stuff and get out of here! You're all fired!"

Farnsworth sat alone in his office, quiet and perplexed while his men filed out. Alternating waves of anger and despair seized him as he tried to assess what McCargar had just done to his life and his work. A unique spirit followed Phil as he walked out of the lab alone that night. Pem already knew what had happened when Phil dragged himself into the house, looking beaten and depressed. She tried to talk about it but Phil was still too over-whelmed and confused to articulate his feelings. They went to a movie instead.

Later that evening, Phil called some of the men to see if they would come back. He reached Cliff first. Cliff, Pem's brother, who had been with them from the very start, reluctantly agreed to come back, but was clearly shaken by McCargar's actions.

The next call went to Tobe Rutherford, who had been a member of the lab gang almost as long as Cliff. "I'm sorry, Phil," Tobe said, "but I'm just not coming back as long as Jess McCargar is part of the picture."

And so it went down the line. One after another, the men of Phil's lab gang said, "Sure, I'll come back, as soon as you do something about Jess McCargar." Phil could not yet fathom how he was going to do that, except for knowing it would take some time and probably another reorganization of the company. But none of these men had the wherewithal to hold out indefinitely. And so over the ensuing weeks, the lab gang slipped away, tak-ing jobs with other companies in the field, such as Dumont, CBS, and even RCA. Some of them might have liked to hold out, but they simply had to take other jobs. Phil slowly realized that it would be impossible to ever get the old gang back together.

Of course, Phil managed to get some semblance of the opera-tion going again over the next few weeks. Starting with the ever loyal Cliff Gardner, he resumed his work, as he had after the previous firings. His work had survived those disruptions, as it had survived the fire in San Francisco that destroyed everything except the spirit that was soldered into his circuits and pumped into his tubes. But that bright, gem-like flame, that vortex which had invented and perfected electronic television, was gone. He had carried on after leaving Philco when he had no idea where the next dollar would come from; he had stood up to the mighty RCA and gone round-for-round with the imposing figure of David Sarnoff. But nothing had quite prepared him for his world to be shattered from within.

Phil did all he could to keep his spirits buoyant while his precious lab gang was absorbed into the fabric of the electronics industry. By now he was familiar with the disappointment, the anger, the frustration, the fear, and the grief that were all part of his chosen profession. But as he returned to the lab each morning in the winter of 1937, something else changed in the way he walked through that space. A new burden weighed on him with every step, and a new emotion took root in the already fragile psyche of Philo T. Farnsworth: In the midst of all his great scientific discoveries, he was learning what it really meant to be a "lone inventor." He had discovered the loneliness.

THIRTEEN

Caught in the Crossfire

On location with a Farnsworth mobile system, ca. 1938

"Man will occasionally stumble over the truth, but most of the time he will pick himself up and continue on."

—Winston Churchill

There was little doubt that by 1937, Farnsworth and his lab gang had truly "turned mankind's most fanciful dream into a startling reality," as the Paramount newsreel had reported the year before. Just looking at the images being sent and received by Farnsworth and his lab gang, or what remained of them in 1937, made it obvious that all of the technological underpinnings that television required were in place. The picture now sparkled with 441 lines per frame, and 30 frames per second. A technique called "interlace" made the image stable, and all the improvements that Farnsworth and his team—as well as engineers from RCA and other laboratories—had added to the system made it ready for public consumption. The technology was ready. The public was more than ready. Only the business institutions that would have to usher the new medium into the marketplace presented any remaining resistance.

As if to underscore television's readiness, the Radio Manufacturers Association—an organization composed of the likes of Zenith, Philco, and of course RCA—established a committee to debate and recommend standards for commercial TV broadcasting. Whatever standard the committee recommended, Farnsworth's system would be able to adopt, and his company would be ready to participate. But first, it was painfully apparent to everybody in Farnsworth's operation that the company would have to face yet another round of reorganization before it would be ready to face the challenges of commercial competition.

As devastating as McCargar's erratic and capricious style of management was on the laboratory's operations, it's not difficult to empathize with the banker's predicament. True, Farnsworth wanted to be in the inventing business, and McCargar seemed to think that they were only in the television business. But after more than ten years of hard work and expense, they still weren't even in *that* business. Patent litigation, spectrum allocation, and the need for signal standards had all conspired to forestall the commercial implementation of the new medium. After a decade of promises and predictions, McCargar, who was getting on in years, was beginning to wonder if his investments in television would ever pay off in his own lifetime.

The emotional and financial stress that accompanied all these issues came to a head during the summer of 1937, when a small group of Farnsworth intimates convened a special meeting in the living room of the Farnsworth's home on Crescham Valley Road outside Philadelphia. In addition to Farnsworth himself, George Everson was present, and brought with him Hugh Knowlton, who represented the prestigious New York investment banking firm of Kuhn, Loeb and Company. Noticeably absent was Jess McCargar.

The meeting was arranged to discuss alternatives to the company's current method of financing its operations, which had re-

remained basically unchanged since 1929, when McCargar superseded Roy Bishop and the Crocker Bank syndicate and began selling stock in the venture to meet expenses. Hugh Knowlton's presence suggested that the current mode of financing was about to change once and for all. Kuhn, Loeb had been involved in some aspects of financing the Farnsworth organization for about a year

Cameraman John Stagnaro

before this meeting; the relationship would prove invaluable as the venture formulated plans for moving forward in the months ahead.

Most conspicuous by his absence from this gathering was Jess McCargar, who was simply not invited to participate—a circumstance that underscores that shift had been happening for some time. Although Farnsworth had been titular head of the company during the thirties, the financial center of gravity for his operation had already shifted to Wall Street. Once Kuhn, Loeb became involved, the firm and its investors began to acquire a significant equity stake in the venture. By the spring of 1937, Jess McCargar was serving at the pleasure of a board of directors that he had not handpicked. In the wake of the summary firing of the entire lab staff, it became evident that McCargar's presence no longer pleased anybody, least of all the interests of Kuhn, Loeb.

This shift in the power behind the company carried even more serious implications for Phil, who was still clinging to the last remnants of his dream that success in television would pave the way for success in whatever line of research he was inspired to follow in the future. But he had also come, reluctantly, to the conclusion that "tacking on shipping room door"—Phil's disdainful metaphor for a manufacturing and merchandising operation—

was inevitable. The course for that new corporate destination was set that day in his living room. If manufacturing would guarantee that Jess McCargar was no longer running the show, then Farnsworth was reluctantly prepared to set up shop.

By dinnertime, the ingredients of a serious proposal had started taking shape. Using the investment banking services of Kuhn, Loeb and Company to arrange the necessary financing, the Farnsworth enterprise would reorganize from the top down. A search would be undertaken to find new executives to run the new company; Kuhn, Loeb would utilize its Wall Street sources to find a factory suitable for acquisition and conversion to, at first, radio and eventually, when the market was opened, to television set manufacturing. The whole venture was to be underwritten by an "initial public offering" of shares on the New York Stock Exchange.

Of course, the newly formed Farnsworth Television and Radio Corporation was not the only company trying to fortify its position in the industry as the final battles over television approached. The list of entries into the television sweepstakes grew longer each day, as companies with names like Dumont, Crosley, Emerson, and Hazeltine, among others, began circling the territory to stake out claims. And, as ever, all the potential players waited to see precisely how RCA would play its hand.

As the growth of the new medium accelerated, the focus of interest began to shift from the financial offices on Wall Street and the research facilities in Camden and Philadelphia to the political corridors of Washington, D.C., where the recently formed Federal Communications Commission was expected to orchestrate the chaos by adjudicating whatever signal standards the Radio Manufacturers Association recommended.

The FCC was moving cautiously, however, realizing that once standards were adopted, the industry would have to live with them for decades, if not centuries, to come. In addition, the FCC was forced to contend with numerous factions opposing the momentum that television was gaining. Radio broadcasters and manufacturers joined forces in an alliance with the movie industry and lobbied to stall TV's progress in Congress and the FCC.

The FCC was also responsible for allocating space for television broadcasts within the electromagnetic spectrum. Because television utilizes extremely high frequencies and much greater bandwidth than sound-only broadcasts, the new medium of television threatened to gobble up an inordinate amount of finite spectrum space; one TV channel would need more spectrum than a dozen radio channels, thus limiting the number of available channels in any geographic region. Still more spectrum space would be required for relay of TV signals if a national TV network were to be established along the lines of existing radio networks.

Without question, the company in the strongest position to capitalize on television was the Radio Corporation of America, whose patent domination in the field of radio transmitting and receiving was almost impregnable. Most observers assumed during the 1930s that RCA would extend its domination to the new field of television, into which company president David Sarnoff had already poured millions and staked his personal legacy. But there was at least one other industrial contender that posed a potential challenge to Sarnoff's vision of television dominance.

The American Telephone and Telegraph Company—the corporate heir to the legacy of Alexander Graham Bell's invention of the telephone—staked out its claim in 1935, when its Bell Labs subsidiary introduced a wired solution to the problem of sending television signals from city to city. This new invention was called a "coaxial cable" owing to the fact that one conductor was threaded through the center of a flexible copper tube. With this development, AT&T placed itself in perfect position to assume the job of wiring together television networks. The FCC tentatively gave AT&T permission to experiment with its cable, and one was strung almost immediately from New York to Philadelphia to begin testing.

In the meantime, the FCC opened an inquiry to make certain that AT&T was not about to create another communications monopoly. Such investigations were familiar territory to AT&T, who, along with RCA, GE, and Westinghouse, had experienced previous governmental inquiries into its affairs. Those earlier inquiries revealed that these giant companies entered into a series of secret agreements during the 1920s upon which the entire structure of the communications business was predicated.

These agreements were ostensibly broad patent cross-licenses. Each company granted the others the use of its vast patent portfolios—with stipulations over how each company could use those patents. Thus, AT&T was able to use all of RCA's patents so long as AT&T stayed out of the radio business, and RCA was able to use all of AT&T's patents so long as RCA stayed out of the telephone business. Conversely, RCA was assured dominance of the radio business as long as AT&T received all the long-lines business from wiring together RCA's radio networks.

These cross-license agreements became notorious during the 1920s and early 1930s as "The Radio Trust." The original cross-licenses were modified—after the threat of antitrust proceedings—in the "consent decree of 1932," which allowed RCA and AT&T to stick to essentially the same terms, albeit couched in less monopolistic terminology.

However, even after the consent decree of 1932 was signed, the agreements omitted one important consideration: television. The AT&T/RCA cross-licenses covered audio transmissions only. There were no provisions at all about which company would use which patents for television. In other words, in a field where everything else was sliced up and nailed down, television was up for grabs. Whoever got there first would call the shots.

Updating the cross-licenses to include television was essential to David Sarnoff's plan for RCA to extend its domination of radio broadcasting into the promising new field. With such patent control at its command, RCA would have a lock on virtually the entire electromagnetic spectrum. To transmit or receive any kind of information via the spectrum would be impossible without employing some sort of RCA-covered device. RCA would truly become THE Radio Corporation and David Sarnoff would become the undisputed Emperor of the Airwaves.

Only one man stood between David Sarnoff and his dreams of an ethereal empire—Philo T. Farnsworth. Sarnoff knew that in order to add television to the existing cross-licenses, each side would have to have patents central to the new art to exchange. AT&T was well prepared to begin negotiating around its contribution, the coaxial cable, and apparently RCA was expected to deliver its end of the bargain in the form of patents that covered the art of sending and receiving video signals. But as things stood in the middle of 1937, RCA didn't own any of those patents. In case after case, the patent office affirmed the fact that the fundamental patents covering the art of television belonged to Philo T. Farnsworth.

RCA and Farnsworth had been litigating over those patents for years, but RCA had little to show for its effort and expense. After nearly fifteen years of consideration, the patent office still had not issued a patent for the Iconoscope, largely because of RCA's insistence that the tube was based on Vladimir Zworykin's 1923 patent application. In numerous decisions the examiners determined that the tube disclosed in 1923 and the Iconoscope were not the same tubes. Furthermore, engineers who had been working with the Iconoscope in the field reported that the Iconoscope was very difficult to use. Its oblique scanning geometry produced a signal that was noisy and needed a lot of filtering. Shading the picture never ceased to be a source of anguish for the first generation of video engineers. The RCA research philosophy—if they couldn't own an invention, they would engineer their way around it—was finally coming back to haunt them.

Sometime early in 1937, some of the men working at the RCA labs in Camden—most notably Harley Iams and Albert Rose—gave their boss something new and exciting he could report to RCA's increasingly impatient board of directors. The engineers had devised an entirely new kind of camera tube capable of producing much sharper, cleaner pictures than either the Iconoscope or the Image Dissector. It employed a novel technique called "low-velocity scanning" (not to be confused with the "velocity modulation" process that paved the way for radar). Sarnoff's legal department informed him that the new development appeared completely original to RCA, and moved quickly to file patents.

Meanwhile, the trademarks department came up with a name for the new tube: it would be called the "Image Orthicon." His confidence restored, David Sarnoff told the RCA board that he had chosen a time and place for launching commercial television service—at the New York World's Fair in April 1939.

Philo Farnsworth found himself caught in the crossfire between two industrial giants in 1937, when the Federal Communications Commission invited him to express his opinions on the future of television. Of particular interest to the FCC at the time was AT&T's apparent failure to issue any licenses for the use of the coaxial cable. It seemed that AT&T was reserving use of the cable for itself, possibly as a bargaining chip in its dealings with RCA, and this would constitute the sort of restraint of trade that the FCC would view rather dimly.

Unknown to Farnsworth as he settled into the witness chair that day in the nation's capital, AT&T president Walter S. Gifford was also sitting quietly in the hearing room. Gifford listened intently as the commissioners began to question Farnsworth about the difficulties he had encountered in obtaining licenses for his patents. Phil told them that he did not hold a license for use of the AT&T coaxial cable, nor had he tried to obtain one.

At this point, Mr. Gifford, hoping to prove a point in front of the Commission, rose to his feet, interrupted the next question, and introduced himself. He then turned to Farnsworth and asked him if he would care to enter into a cross-license agreement with AT&T. Farnsworth, a bit stunned, responded that of course he would welcome an exchange of patents with AT&T.

"Then see me after you are through here," Gifford said, and sat down, leaving the rest of the room in silence.

Unlike RCA, which had staked its entire future on developing electronic television on its own, AT&T had nothing to lose by offering Farnsworth a cross-license, and everything to gain. By making the offer in the presence of a commission that was investigating the company's business practices, Gifford seemed to contradict any charges of monopoly, and his clever staging cast AT&T as the champion of free enterprise. More important, by licensing the Farnsworth patents, AT&T was firing a clear shot across RCA's bow, loudly announcing their intent to do battle in the new arena.

It took more than six months for AT&T and Farnsworth to iron out the specifics of their cross-license agreement. However, the deal ultimately worked out was uniquely beneficial for the Farnsworth camp. The cross-license gave each company equal access to the other's patents, but AT&T also agreed to pay a substantial royalty to Farnsworth, without requiring that Farnsworth pay any reciprocating royalties to AT&T. When the papers were signed on July 22, 1937, Farnsworth was so excited at the prospect of this important recognition that he was barely able to "affix a legible, though squiggly 'Philo T. Farnsworth' to the important document."[81]

If David Sarnoff read the August 14 issue of *Business Week*, he received a rude awakening: The AT&T-Farnsworth deal meant "that the grip which the Radio Corporation was generally assumed to have on the future of television was relaxed," because "Farnsworth now obtains access to the basic broadcasting patents. In other words, he is now able to compete with RCA on more equal terms. The road is no longer blocked should Farnsworth decide to enter manufacturing."[82] The license with AT&T could not have been more timely, because entering manufacturing was precisely what Farnsworth's new backers were preparing to do.

In effect, Farnsworth had sneaked in through the back door and raided Sarnoff's kingdom. If AT&T and Farnsworth wanted to, they could have launched commercial television themselves, leaving RCA to peer in from the outside with its cranky Iconoscopes.

The Farnsworth deal with AT&T had immediate implications for the rest of the industry, which greeted the news as a major coup for Farnsworth. For the first time, other companies could purchase equipment from Farnsworth without fear that RCA would interfere with the transaction or threaten their licenses. The Columbia Broadcasting System was one of the first companies to buy Dissector tubes from Farnsworth for TV experiments they were conducting from the Chrysler Building in New York City.

All this activity was a clear signal to Sarnoff that the industry was beginning to accept Philo T. Farnsworth on his own terms. Still, Sarnoff continued to hold his cards close, confident that the new Image Orthicon would provide his ace in the hole. Further testing with the new device confirmed earlier indications that the Orthicon outperformed all its predecessors. Almost the moment the Image Orthicon was proven, RCA junked its research on the Iconoscope with quickness that suggests that the only reason the Iconoscope was ever supported by RCA was to get around Farnsworth's patents.

Sarnoff did not worry too much about the AT&T-Farnsworth deal until the day his lawyers informed him that their detailed patent search had encountered U.S. patent #2,087,633.

The device disclosed in this patent was labeled an "Image Dissector," but it employed a low-velocity scanning process—the very same process that made the Image Orthicon such an efficient and reliable camera tube. Of course, the patent had been issued to Philo T. Farnsworth—in 1933, four years before Rose and Iams demonstrated the first of their Image Orthicon designs for RCA.

For Farnsworth, the Image Orthicon was another manifestation of his basic philosophy, that being first would lead his team into new territory and produce ever more valuable inventions. For RCA, the Orthicon represented one more obstacle in their boss's grand design. Once again, RCA's legal forces mobilized to open interference proceedings with the patent office over the conflicting claims, but this effort met the same result as every previous RCA challenge. Priority on claims relating to the low-velocity scanning camera tube was awarded to Farnsworth. The only ingredient on which the patent office would award ownership to RCA was the name itself, "Image Orthicon," which RCA had registered separately as a trademark. So, the Orthicon tube, the workhorse upon which the television industry was built in the forties and fifties, was basically a Farnsworth invention wearing an RCA name.[83]

By 1938 David Sarnoff had spent more than $10 million for research in television, yet RCA was unable to secure but a handful of patents, none of them as essential to the art as those belonging to Farnsworth. On the other hand, Farnsworth's research had cost less than $1 million, and his portfolio controlled fundamental aspects of the new technology. Sarnoff was outspending Farnsworth by ten to one, and had virtually nothing to show for it.

When Roy Bishop said, back in 1928, it would take "a pile of money as high as Telegraph Hill" to bring television to the marketplace, he was right in at least one regard: it did take a huge pile of money. Only it was RCA's money, most of which was spent trying to circumvent the Farnsworth patents. Such is the difference between trying to engineer an invention and actually inventing one.

About this time, Sarnoff began to change his tune. In public addresses, he mentioned Farnsworth several times by name and acknowledged his contributions to the art of television, although he was always mentioned in the company of other inventors, particularly Zworykin. Nevertheless, as Farnsworth's people learned about such utterances, their conclusion was confirmed through the Wall Street grapevine: RCA's change of heart was real. So in the closing months of 1938, lawyers for Farnsworth and RCA exchanged the first overtures on the subject of the long-awaited patent license.

FOURTEEN

Gone Fishin'

The Farnsworth's property in Brownfield, Maine

"The Universe is not only stranger than we imagine,
it is stranger than we *can* imagine."
—Arthur C. Clarke

In the months after the pivotal meeting in the Farnsworth living room, Hugh Knowlton of Kuhn, Loeb and Company went to work, utilizing the resources of one of Wall Street's most prestigious investment banking houses for the organization of the new Farnsworth Television and Radio Corporation. Farnsworth accepted, however reluctantly, the decision to pursue manufacturing, hoping that the presence of a firm like Kuhn, Loeb would give his company a well-financed foundation on which to build its future solvency. George Everson and other officers of the company spent a good deal of their time in New York, working closely with Kuhn, Loeb and other Wall Street contacts, while Phil monitored the progress from the lab in Philadelphia.

Between his weakened health and shattered lab gang, it was difficult for Phil to stay focused while the real future of the business was being determined elsewhere. He continued to work closely with Cliff Gardner and the skeletal staff that he still had around him, suggesting further refinements to their television system. But the more he thought about what he had already accomplished, and the more he looked at the picture his tubes produced every day, the more apparent it became to him: the "inventing" part of television was pretty much done. While there would always be a dimension of novelty to further engineering, and while patents would be filed and contested in a never-ending process of technical evolution, the fundamental aspects of inventing the new medium had become mostly a matter of history.

In the crucible of his laboratory over the preceding decade, Phil Farnsworth had learned as much as any of his contemporaries. His numerous inventions and patents were the result of a philosophy that governed his work every day: "Let's see what we can learn in the lab today." After more than ten years of approaching his work with this attitude, he had learned as much about how electrons behave and how they can be manipulated as anybody operating in the field at the time. All those lessons now suggested their own avenues of further exploration, and Farnsworth began to indulge his imagination in the more rarefied atmosphere of conceptual research, in which an invention is not so much an end in itself as it is a way of proving something, in order to advance the theoretical base of scientific knowledge.

In his daily work, Farnsworth was often reminded of things that had kindled his boyhood interest in the theories of Albert Einstein, and he found a fair amount of his mental energies thinking about the possibilities that lurked in those conceptual areas. From somewhere deep in the river of knowledge and discovery, a Siren voice was calling to Philo Farnsworth.

In Jess McCargar's last official act as president of the company, he summarized the condition of the company in a memorandum addressed to "The Stockholders of Farnsworth Television, Incorporated" dated August 9, 1937.[84] The primary reason for the memo was to announce the license with AT&T, which indeed was momentous news for the stockholders. The AT&T license told them in unequivocal terms that their company had arrived, that their speculative faith in Farnsworth and his invention was about to pay off. These stockholders, through their investments, had helped to transform a struggling start-up into a substantial player with a significant role to play in the impending launch of a whole new industry. The license, McCargar wrote, "constitutes a recognition of our patents by the greatest corporation in the world."

In a challenging nod toward RCA, the memo further stated that the AT&T license "also insures that with the advent of commercial television, there will be a source of television transmitters [the Western Electric Company] which will be able to supply the demand." McCargar hinted at further progress, adding that "several tentative negotiations, looking toward the granting of further license by us which will speed television progress, are now going on." The implication was clear: RCA had better be willing to come to the table soon or there might not be a chair for them.

The memo went on to describe technical progress in the Mermaid Lane laboratory, the success of experimental broadcasts from the Wyndmoor studio, and the stunning success of Farnsworth's Multipactor tubes. It cites "one hundred and thirty-five applications for United States patents have been filed...of which thirty-three have matured into patents, six have been allowed but have not yet been issued, six have been abandoned, and the remainder are pending."

The memo also described the progress of various patent interferences, reporting a number of which "have been finally settled," most notably the cases "wherein it was determined that Zworykin could not lay claim to the Farnsworth Dissector tube." Finally, McCargar assured the stockholders that "our own television patent structure is stronger than any other individual structure in this field. With the American Telephone and Telegraph patents, we believe the combination is now the strongest in existence."

The tone of the memo is very straightforward and businesslike. McCargar sounded almost diplomatic. There was no hint of the contention and rancor behind the scenes. Rather, the letter offered a nice summary of exactly where the company stood in the late summer of 1937, almost exactly ten years from the day that Farnsworth transmitted his first straight line image. And when he signed for the "Board of Directors, Farnsworth Television Incorporated, J.B. McCargar, President" there was no mention of the fact that this is the last such letter that Jess McCargar would ever sign.

By the time this memo was sent to the stockholders, control of the company was already in the hands of Hugh Knowlton and his associates at Kuhn, Loeb. George Everson remained in the inner circle, as George was the one with a long-standing personal relationship with Knowlton, who had first brought Farnsworth's work to the attention of the investment firm. Now, Knowlton and his colleagues went about the serious business of creating a credible corporation out of Farnsworth's work, one that would be taken seriously by Wall Street.

One major challenge they faced was finding a suitable chief executive. The search produced Edward A. Nicholas—who liked to be called "Nick." At the time Nick was approached about the top Farnsworth post, he just happened to be the vice president of the licensing division of RCA.

Nicholas's rise through the ranks of RCA was not unlike that of his boss, David Sarnoff. Like Sarnoff, Nicholas had started as a messenger boy for a wireless company, eventually becoming a wireless operator himself. Also, like Sarnoff, Nick had found work with the American Marconi company, and in the 1920s he had even served a brief stint as Sarnoff's personal assistant. In addition to his experience with engineering and patent matters, Nicholas had a background in manufacturing and distribution.[85] Though he was lucratively employed and enjoying a promising career with RCA, his current employer, it is not surprising that somebody in Nicholas's position would change horses in favor of the top job with a risky new company; the new position offered opportunities that could never exist at RCA so long as David Sarnoff remained in command. Nicholas relished the challenge, since working with Farnsworth presented him with an opportunity to build an enterprise that one day could conceivably rival his former employer.

After extensive background checks, Nicholas was offered the position of president and CEO of Farnsworth's reorganized company when it was ready to begin operations. The Board also approved the appointment of Edward Martin, from the Hazeltine Electronics Corporation, to serve as corporate counsel. Both men waited in the wings while the Wall Street operatives continued assembling the rest of the new corporate puzzle.

The deal began to crystallize when Kuhn, Loeb learned of a plant in Fort Wayne, Indiana, that was being sold as part of the liquidation of the Capehart Company, once regarded as the most elegant name in automatic record changers and jukeboxes. The company had fallen on hard times, and was finally forced into bankruptcy when one of its most expensive mechanisms developed a habit of breaking the records it was supposed to be gently changing. Aside from that minor detail, the company's plant was an ideal facility, and the name "Capehart" was expected to lend a certain cachet to the eventual Farnsworth product line.

With a management team ready to take residence in the executive suite and a suitable factory ready for acquisition, Kuhn, Loeb went about the task of organizing the stock offering. This turned out to be a far more arduous endeavor than anybody anticipated, as Wall Street in general was just learning its way through the Byzantine policies and procedures of the newly formed Securities and Exchange Commission. The SEC had been created during the Roosevelt administration to prevent capital market abuses, which some economists and politicians believed had caused the Great Depression. The consequence was a mountain of paperwork that confronted any company wishing to sell its shares to the public. The Farnsworth filing with the SEC required many months and more than 50,000 in lawyers' fees. As George Everson described the process, "Since it was a perfectly legitimate speculative undertaking, it seemed to us that there should not be a great deal of red tape involved. However, in our case, the attitude seemed to be one of daring us to prove that we were not a pack of dishonest rogues."[86]

Though the primary focus of his thoughts remained out on the unexplored frontier that kept calling for his attention, Farnsworth continued to exhibit the sort of work habits that nowadays would earn him the label of "workaholic." He continued to spend part of every day at the laboratory, which filed another nineteen patents during 1937 and 1938. And he continued to serve as resident celebrity at the studio hosting visiting dignitaries. But these activities no longer energized him as they did before the lab gang was fired.

Nor was he drawing much inspiration from the machinations in New York, though he took an active role in the process, often catching the train to the city to work on some aspect of the preparation. After his experiences with McCargar, Farnsworth was very concerned that this time around things be organized more constructively. While the Wall Street contingent had a large voice in the reorganization, Phil still had clout: if he turned "thumbs down" on some aspect of the deliberations, that was usually enough to end that particular initiative. But he didn't use his clout very often. Despite his strong personal interest in the fate of the emerging company, Phil most often erred in favor of generosity rather than negativity. He was more inclined to go along with things, to let issues demonstrate their own workability or lack thereof. And although he expressed some reservations about the hiring of Ed Martin as corporate counsel, he had more confidence in Nick Nicholas and thought that his new CEO might indeed be the executive antidote to Jess McCargar.

Still, there was not much inspiration for Phil at the lab, and there certainly wasn't much among the bankers and businessmen in New York. For real inspiration, Phil turned within himself and challenged his imagination to push back his own mental frontiers. He was finding this new territory rich and fertile. But all Pem could see, in the spring of 1938, was that Phil seemed to be losing his strength again. In fact, it seemed to her that he had never fully regained his health in the difficult days after their return from Europe.

Fortunately for Pem, her brother Cliff had developed over the years into what the family regarded as a "high-brow" fly fisherman, an artist with a rod and light-test line who tied his own flies and knew the best fishing spots all over the eastern seaboard. Thinking a change of scenery would do her husband some good, Pem suggested that she and Phil join Cliff and his wife, Lola, for a fishing expedition to Florida. Phil, tireless worker that he was, was reluctant to spend so much of his time on the end of a fishing pole. Nonetheless, realizing that there was not much to do until reorganization was finished and the lab was relocated, he agreed to take off and spend time with his wife and family.

The foursome made it only as far as Snowbird Creek in North Carolina, but that was all Phil needed to discover, to his pleasant surprise, that fishing put him in just the right frame of mind, presenting him with an unhampered opportunity to let his mind wander wherever it wanted to go. The end of a fishing pole, it turned out, was every bit as good a place to think as any other, maybe better. By summer he was spending as much time fishing as he was anything else. For the first time, he could feel his internal compass telling him that it was time to turn his energies in an entirely different direction.

After another fishing trip to the northern reaches of the Appalachian Trail, Phil and Pem stopped in Brownfield, Maine, to look in on a property that George Everson had acquired in a foreclosure deal during the Depression. Pem gasped when she first saw the place. The house itself had been built during the colonial period, but was sorely neglected. "People had come and taken all the doors and the windows and the deer and the porcupines had taken over," Pem remembered, but "the old house was strongly built from hand-hewn beams—you could see the axe marks in the wood—and there was beautiful pumpkin pine paneling. Otherwise, the house was nothing."

Pem didn't realize the importance of finding a saving grace in this run-down old farmhouse until she and Phil explored the rest of the property. "We walked down through the fields, and Phil sees this little spring-fed stream and a swale and he says, 'I'm going to buy this place and build a dam there. I'm going to have a trout pond.'" At that point Pem must have realized that the dilapidated old farmhouse that provided shelter for the porcupines and deer would soon be her home.

Phil wasted no time burning up the wires to San Francisco, asking George to sell enough of his stock so that he could buy the eighty-acre farm. George was less than thrilled at the prospect of selling stock to cover a bit of Phil's personal business, particularly in the midst of all the other critical business that was crawling along in New York, but somehow he managed to accommodate Phil's request and made the necessary arrangements to sell him the farm.

Pem and Phil prepared to spend the following spring and summer at their new Maine homestead. Their family now included two sons: Philo III, now nearly ten years old, and little Russell Seymour, who had been born in October 1935 and nicknamed "Skee" like his namesake, Skee Turner. The four Farnsworths, along with Pem's father, spent the summer of 1938 camping in the remains of the old farmhouse. Phil supervised the pouring of the concrete core for his trout-pond dam, but in the back of his mind, he was building the nest that would nurture and hatch his next Big Idea.

Meanwhile, back in New York, the wheels of capitalism continued to grind in their slow, inexorable way. The filings and counter-filings with the SEC for the initial public offering, and the negotiations for the acquisition of the Capehart factories in Indiana continued through the fall of 1938, and into the winter of 1939. By the time the process neared its conclusion, it had taken nearly two years to hammer out the details of the various deals that would put Farnsworth into the electronics manufacturing business.

All the contracts, agreements, and promissory notes that would finalize the plans first outlined in Farnsworth's living room in the spring of 1937 were finally ready to be ratified in March of 1939. Among other things, the papers included provisions for an initial public offering of three million dollars-worth of stock that would capitalize the new Farnsworth Television and Radio Corporation.

Tensions mounted when the closing was held up for two additional weeks, while the Wall Street contingent waited for a weak market to strengthen before floating their issue. Finally, on March 14, 1939, everything was ready, the market was firm, and the deal was closed. When the documents were all signed, George was handed a check for $3,000,000, and the Farnsworth Television and Radio Corporation was open for business.

The following day Hitler invaded Czechoslovakia.

FIFTEEN

Loggerheads

In 1939 Philo T. Farnsworth was named
"One of America's Top Ten Young Men"

"To live a creative life, we must first lose
our fear of being wrong."
—Joseph Chilton Pearce

Phil and Pem stayed in Maine through the spring and summer of 1939 while the Philadelphia lab was crated up to be hauled to Indiana. They expected to take up residence in Fort Wayne before the next winter. Meanwhile, new pages in the history of television were being written every day, and public interest in the imminent arrival of the new medium continued to intensify.

With a flourish designed both to capitalize on the public's curiosity and to lend historical stature to the event, David Sarnoff arrived at the opening of the New York World's Fair on April 30, 1939, with an entourage that included Franklin D. Roosevelt. FDR became the first President of the United States to appear on television in a ceremony carefully staged for the benefit of RCA's television cameras. In his opening remarks, Sarnoff announced the arrival of a new epoch:

"And now we add radio sight to sound. It is with a
feeling of humbleness that I come to this moment of
announcing the birth in this country of a new art so
important in its implications that it is bound to affect all
society. It is an art which shines like a torch of hope in
a troubled world. It is a creative force which we must
learn to utilize for the benefit of all mankind."

There was no mention of any of the individuals who were directly responsible for that groundbreaking accomplishment. Not
even Zworykin earned a word of Sarnoff's praise, and certainly
not Farnsworth. No, this was entirely Sarnoff's event, his carefully
orchestrated testament to the power of corporate engineering,
and the opening salvo in his PR campaign to establish himself as
"the Father of Television." There was no room for any individual
achievement other than Sarnoff's own. This "birth of a new art,"
he would have us believe, was entirely his own doing as the driving force behind the engineers at RCA. Individuals, least of all
individual inventors, need not apply.

The event was televised to an audience on the fairgrounds, and
broadcast to a handful of receivers in the New York area. Later that
week, television receivers went on sale in limited quantities at department stores in New York, at prices ranging from $199 to $600
(approximately $2,000 to $6,000 in 2002 U.S. dollars). These first
commercial sets used the 441-line/30-frame standard proposed to
the FCC by the Radio Manufacturers Association. The Commission
had still not established official industry-wide standards but that
did not stop other companies from announcing that they, too,
would soon be selling TV receivers. The industry was eager to follow RCA's plunge. Ignoring the FCC's delay on formalization of
signal standards, they also overlooked the FCC's denial of licenses
for anything other than the experimental use of television.

Not even the mighty RCA had permission to sell commercial
time to advertisers to support television broadcasting. Sarnoff
wanted the World's Fair opening to go down in history as the
arrival date for commercial television (that purpose was accom-

plished, as "Golden Anniversary" observations in 1989 attest). However, knowledgeable observers at the time regarded the event in more sanguine terms; rather than opening the market for commercial television, the event only signaled the beginning of another phase of experimentation, one in which the public would be allowed to participate through the availability of a handful of receivers that were sold at retail. Television's commercial payoff was still years away.

In a May 1939 issue, *Fortune* magazine followed up the World's Fair event with a broad assessment of television, hailing the new medium as Sarnoff's "Thirteen Million Dollar 'IF.'"[87] Although Sarnoff omitted any mention of Zworykin at the World's Fair, the *Fortune* article was less neglectful; it appended Zworykin's legacy to Sarnoff's, becoming one of the first widely distributed publications to recite the dubious litany of the Iconoscope. Stating that it was "first patented in 1923," *Fortune* noted that "the tube was not ready for practical demonstration until ten years later."[88] There was no mention, of course, of the patent office's ruling in 1934 that the Iconoscope and the tube disclosed in 1923 could not have been the same device.

The *Fortune* article was more generous to Farnsworth in some respects, acknowledging that "most of the laboratory work has been done by large companies...with the noteworthy exception of Philo T. Farnsworth."[89] Though the article was primarily deferential to RCA, it did hint at the unresolved issues that continued to cloud the future of television:

> Until the patent office and the courts unwind the snarl of claims and counterclaims, no one will know the exact patent structure of the television industry. It is certain...that Farnsworth and RCA together will hold a large majority of the worthwhile television patents. Because of RCA's radio patents, which are necessary in the manufacture of any television equipment, no one will be able to enter the field without an RCA license. Whether only an RCA li-

cense is needed remains to be seen, but the point will probably not arise, since a cross license agreement between RCA and Farnsworth is imminent.[90]

What the Fortune article failed to point out, however, was that Farnsworth, by virtue of his cross-license with AT&T, already had access to the RCA radio patents. The real question, then, was whether RCA would ever get access to the Farnsworth patents. Even as Sarnoff gambled with his "launch" of television at the World's Fair, RCA and Farnsworth were still at loggerheads in their negotiations for a patent license that would enable RCA to put its market muscle into the new industry and finally deliver television to the masses.

With the new company, Farnsworth Television and Radio Corporation of Fort Wayne, Indiana, now in place, Nick Nicholas, the new company's CEO, and his corporate counsel, Ed Martin, stepped in to handle the negotiations with RCA. This was familiar territory for both men. In his previous position, Nicholas had negotiated patent agreements on behalf of RCA; Martin, in his position with Hazeltine, had often been on the opposite side of the table from Nicholas. Now both men sat on the same side of the table representing the principal inventor of electronic television, and Otto Schairer sat on the other as chief patent counsel for RCA.

RCA was willing to concede that it would not be possible for it to legally manufacture electronic video components—neither cameras, transmitters, nor receivers—without employing techniques covered by Farnsworth's patents. The Farnsworth portfolio included all phases of electronic scanning and synchronization, electrostatic and magnetic focusing, electron multiplication, the sawtooth wave, blacker-than-black horizontal blanking—in short, all the fundamentals of manipulating electrons to send pictures through the air. Farnsworth's priority had been repeatedly affirmed by the patent office, but Otto Schairer persisted with his intransigence, demanding that the proposed license be entirely reciprocal; RCA still refused to pay Farnsworth royalties for the use of his patents.

The negotiations bogged down when RCA proposed a clever variation of its now-familiar trading philosophy. Still unwilling to pay a continuing royalty, RCA proposed to pay a fixed sum in advance, insisting on a rather meager figure, something in the mid six-figures. Farnsworth's representatives knew

David Sarnoff opens the RCA pavilion at the 1939 New York World's Fair

their patents were worth many times that amount. Farnsworth's representatives flatly rejected the proposal and RCA went back to the drawing boards.

While the fate of his patents was being hammered out in New York, Farnsworth returned to his retreat in the woods of Maine with the spring thaw. The spring and summer of 1939 were spent restoring the old farmhouse, making it livable after decades of neglect. The windows and doors were replaced, the fireplaces restored, and the kitchen rebuilt into an updated variation of its original configuration, complete with a brick baking oven.

The big project that summer was rebuilding the core for the trout-pond dam on the stream that ran through the property. The first dam, erected the previous fall, was washed out during the winter, so this time Phil resolved to put a much sturdier structure in its place. He wanted to anchor the new dam in bedrock, but after digging to a depth of thirty feet found only compacted glacial silt. So he redesigned the new structure with twenty-foot wide wing walls and a twenty-five-foot high foundation sunk twelve feet into the ground. The whole structure was laced with reinforcing steel rods, and the concrete was poured in a marathon "continuous pour" that lasted more than thirty hours and required the help of almost every able-bodied man in or near the village of Brownfield. When Phil Farnsworth was determined, mere engineering could not deter him; by the end of the summer the dam was finished, and Farnsworth had his well-stocked, private trout pond.[91]

Shortly after the family's arrival in Maine for the summer, Farnsworth learned that he had been named one of "America's Top Ten Young Men" in an annual ritual performed by biographer Durward Howard. This confirmed Phil's stature in the pantheon of 20th century luminaries. Among the other

Pouring the dam for Farnsworth's trout pond in the spring of 1939

personalities so anointed for 1939 were no less than Lou Gehrig, "baseball's iron horse"; Earnest O. Lawrence, creator of the cyclotron "atom-smasher"; William S. Paley, founder of the Columbia Broadcasting System; Harold Stassen, America's youngest governor (of Minnesota); and Spencer Tracy, two-time Academy Award–winning Hollywood actor. Most of the men who shared this honor remain famous to this day, but Philo Farnsworth was already retreating into a self-imposed form of obscurity.

Though the farm in Maine was entirely habitable by the end of the summer, Phil, Pem, and the boys proceeded to set up housekeeping in Fort Wayne, where Phil assumed his new position as vice president and director of research for the Farnsworth Television and Radio Corporation. Despite his own misgivings about assuming such a role, Farnsworth became actively involved in assembly-line engineering and product design.

It wasn't long, however, before his mind went back to the subjects that had preoccupied him earlier. Product engineering became tedious and boring, but he considered it part of his personal obligation to finish what he had started. So he spent most of his days working with engineers at the plant and his evenings working over the ideas and equations that truly challenged his imagination.

SIXTEEN

False Dawn

Employees of the Farnsworth Television & Radio Corporation

"The most interesting phenomena are of course in
the new places, the places where rules do not
work—not the places where they do work! That
is the way in which we discover new rules."
—Richard Feynman

The final chapter in the invention of television was written in
September 1939 in a conference room high above the side-
walks of New York's Rockefeller Center, where a handful of rela-
tive strangers assembled to finalize the long-awaited cross-license
between RCA and Farnsworth. Ironically, not one of the principals in the story was present. Neither David Sarnoff, Vladimir
Zworykin, nor Philo T. Farnsworth were in attendance for this
epic gathering. In their place gathered Nick Nicholas and Ed Martin representing the Farnsworth interests, and Otto Schairer and
RCA's corporate secretary, Lewis MacConnach.

After Nicholas and Martin signed the papers, Otto Schairer sat down to affix his own signature to the historic agreement. When he was done, the Radio Corporation of America, which heretofore only *collected* patent royalties, had agreed for the first time in its corporate history to *pay* the Farnsworth Television and Radio Corporation a minimum of $1,000,000 over the course of ten years.

Legend has it that Mr. Schairer had tears in his eyes as he signed the document.[92]

Although the agreement was announced in a press release on October 2, 1939,[93] the occasion was accompanied by very little fanfare, passing virtually unnoticed except within the industry. Understandably, RCA was not eager to publicize the terms of the agreement, lest the industry get the impression that a patent royalty window was now open at RCA. To the contrary, RCA's capitulation to Farnsworth strengthened Sarnoff's resolve that such a license would never happen again.[94]

The Farnsworth family remained in Fort Wayne through the winter of 1939 to the spring of 1940. Phil found the work environment increasingly stifling, but stayed on "for the good of the company."[95] Indeed, in its first few years in Fort Wayne, the company prospered, as Nicholas negotiated numerous distribution deals to sell Farnsworth and Capehart-Farnsworth branded radio sets in retail outlets throughout the country. Phil was particularly pleased when Nick Nicholas acquired one of the local Fort Wayne radio stations, WGL, to use as a launchpad for a Farnsworth owned and operated television network. With this initiative, Nicholas seemed to concur with Phil's own oft-stated contention that broadcasting would ultimately prove to be a far more lucrative business than manufacturing.

But if television was just around the corner—as David Sarnoff started saying back in 1936—then that corner turned out to be World War II. Just two weeks prior to the agreement with RCA, Hitler had invaded Poland, an act of aggression that dragged all of Europe toward a precipice. It seemed only a matter of time before the United States was also drawn into the conflict.

Despite the threatening developments overseas, the industry proceeded with its efforts to settle on signal standards, as if its leaders were in a race to get television established before the country found itself at war. Attention turned to the FCC, which in the winter of 1940 agreed to consider the RMA's recommendation to adopt the 441-line standard—the same format that RCA had employed at the World's Fair and that other manufacturers were also now advocating. Phil, and Ed Martin attended a number of hearings and joined forces with RCA in lobbying for the 441-line standard, as both companies were now equally anxious to get television out of the gate in order to capitalize on their investments.

The FCC moved very cautiously. The Commission knew all too well how rapidly the technology of television was evolving, and was reluctant to lock the industry into a standard that did not truly reflect the state of the art at that time.

The FCC finally adopted television signal standards in May 1941. The good news for the industry was that it finally had a single set of technical specifications around which to begin manufacturing and broadcasting. The bad news was that instead of the 441-line standard that both RCA and Farnsworth had been pushing for, the Commission settled on a standard requiring 525 scan lines per frame. This was a blow to any consumer who had purchased a television set in the months immediately following the World's Fair; those systems employed the 441-line standard and were suddenly rendered obsolete. The decision was also a blow for those manufacturers who were already selling sets—most notably RCA—who would now have to retool their assembly lines to conform to the new standard. Fortunately, the Farnsworth plants had not yet started manufacturing television sets.

Nevertheless, all of the discussion of signal standards and assembly lines quickly became irrelevant. Later that same month, President Roosevelt declared a national state of emergency and ordered that any raw materials that could be applied to the manufacture of television or radio sets be diverted to the production of military hardware.

Thus, the period between the World's Fair and Pearl Harbor came to be known as the "False Dawn" of television.[96]

For Philo T. Farnsworth, it was more like a premature sunset. Once again, forces beyond his control conspired against his health and well-being. He was not only frustrated in his efforts to bring commercial television to fruition, but also discouraged as he began to realize that the years of delay had eroded most of the effective period of his patents. Only seven years remained before his first patents, issued in 1930, would expire and enter the public domain. These were the patents containing the historic and controlling language about the "electrical image," which embodied the art most essential to the creation and transmission of electronic television images. If the coming war lasted as long as Farnsworth suspected it would, he feared he might never earn any royalties from those patents.

This fear, along with the tiresome grind of his work in Fort Wayne, the national emergency, diversion of raw materials, and the ominous specter of war would have been enough to send any man down the slippery slope of depression. Indeed, the bubble that Farnsworth had kept inflated by the sheer force of his brilliance, the dream he had carried on the shoulders of his will, all seemed to come crashing down around him.

Besides all the external circumstances that were out of his control, Phil's own brilliant mind, which he had trained over the years to work even while he slept, was beginning to turn on him.

At one point, the usually composed Phil broke down and told Pem, "I just can't go on like this anymore. I guess I've trained my brain too well; now I have a hard time turning it off. I'm afraid it's come to the point of choosing whether I want to be a drunk or go crazy."[97]

Alarmed, Pem suggested Phil consult a physician, perhaps even a psychiatrist, but Phil scoffed at her suggestions. He'd seen doctors in the past and they hadn't been much help. "Had they known then what they do now about treating depression," Pem reflected later, "they could have saved us both years of pain and anguish."[98]

Not uncharacteristically, Phil had his own ideas about what sort of therapy he needed. He called Nick Nicholas and told him he was going to take a few weeks off. Then he tried to drink himself into oblivion. He drank day and night, to the point that Pem seriously threatened to leave him. Somehow her threat got through his alcoholic haze, and he agreed to see a doctor recommended by colleagues at the plant. The doctor made a house call and prescribed what Pem called "an elixir," with the assurance that it was "harmless and non-habit-forming." Phil responded to the potion, but only for a few days, after which he decided on his own to double his dosage. He confined himself to bed and refused to eat.

When Hugh Knowlton learned how precarious Phil's situation had become, he recommended a specialist he knew in Boston. Pem agreed to take Phil to this doctor—but first she had to get Phil on a train. A close family friend helped Pem by tenderly carrying Phil's frail, emaciated frame "like a baby" onto the train, and they managed to make their way to a hospital in Boston. The doctors there examined Phil and informed Pem that the potion Phil was taking was "chloral hydrate—a highly addictive poison."[99]

Phil responded to the treatment in Boston, and after four weeks in the hospital was again enough of his impatient self to insist that his recovery continue on to the farm in Maine. With adequate provision for his continued care, the doctors agreed, and the family returned to its Maine homestead. Once there, Phil found renewed strength of both body and spirit, surrounded by his family and the piney woods of Maine.

When he learned that the FCC had indeed suspended the approval of licenses for commercial television broadcasting indefinitely, he put television out of his mind, seemingly for good, and began making plans for the property. He arranged to have some equipment and his journals shipped from Fort Wayne, and laid the foundation for new a three-story structure that would serve as a private laboratory. With the war inevitable and little happening in Fort Wayne, the Farnsworths decided to "settle in for the duration."[100]

Comfortably ensconced in his wooded hideaway, Farnsworth found his own way to contribute to the war effort. When he learned that a local lumber mill was planning to clear cut thousands of acres of timber adjacent to his property, Farnsworth acquired the company. Turning the daily management over to his brothers, Carl and Lincoln, he implemented a program of selective cutting that would not damage the landscape nor disturb the stirring vistas from his living room. Soon, with carefully controlled lumber harvesting, the Farnsworth Wood Products Company was delivering a prodigious amount of milled wood—two boxcars each week—to the War Production Board, which made boxes for bullets to ship to the fronts in Europe and the Pacific.[101]

Back in Fort Wayne, the Farnsworth Television and Radio Corporation threw itself into the war effort with similar fervor, producing military communications and radar equipment worth more than $100 million over the course of the war. Nick Nicholas reluctantly granted Farnsworth a leave of absence, and agreed to provide some funding for his new lab so that Phil could both stay involved in the company's war effort and provide the staff with ideas for new products that would keep the lab in Fort Wayne humming. Phil invited Cliff Gardner and two other staffers from the Fort Wayne lab to join him in Maine. With Cliff at his side making tubes again, Phil reclaimed a sense of the camaraderie of the early days.

One of the first tasks Phil assigned to Cliff was to fabricate a Klystron tube, the device that other workers had patented around the electron-bunching technique that Phil had pioneered back in 1936. The Klystron was a key component of radar. Knowing that radar would be instrumental in the war effort, Phil wanted to see what else he could contribute. When the Cliff's first Klystron was activated, Phil got more than he'd bargained for: there, in the center of the Klystron, was that same mysterious blue glow he had observed earlier in the Multipactor tube. He was determined to investigate the phenomenon further, when his attention was diverted by another pressing matter.

The Farnsworth property in Maine during World War II

Not long after the bombing of Pearl Harbor, Farnsworth received a visit from Dr. David Webster, a professor of physics at Stanford University who flew his own small airplane in short hops all the way across the country to Maine with an intriguing, mysterious offer. Dr. Webster wanted Phil to join an important "government research project" in Chicago. When Phil tried to learn more about the project, his visitor refused to say any more. The project was top secret. But Phil had his suspicions.

"I think they're trying to build an atomic bomb," Phil confided to Pem, "but I don't want anything to do with it."[102] Farnsworth was quite comfortable making boxes for bullets, but he would have nothing to do with using the power of his knowledge to build a weapon that could destroy entire cities.

The project that Webster tried to recruit Farnsworth for turned out to be Enrico Fermi's first "atomic pile." On December 2, 1942, the Italian physicist Fermi and his team, working in a converted squash court beneath the bleachers of a football stadium in Chicago, succeeded in achieving the first controlled release of the energy that Albert Einstein had long predicted would be found binding the nuclei of atoms. Fermi's Chicago Pile, indeed, turned out to be an important breakthrough on the path that only three years later would deliver the first atomic bombs.

Farnsworth could not have known precisely what Dr. Webster was proposing, nor what Fermi was working on in Chicago. But it is clear that Farnsworth's creative processes were taking him down a similar path, as he contemplated his own ideas about how the energy that Einstein discovered might be released from atomic nuclei. Although his own work over the previous decade and a half had focused primarily on the electron—the particle that orbits around the outer layer of atoms—his awareness never strayed far from those forces and particles that lay buried within the inner layer, the nucleus. Indeed, such notions had been tugging at the back of his mind since his earliest encounters as a boy with the theories of Einstein.

For a while, at least, Farnsworth's lab in Maine was precisely the sort of scientific Shangri-la that he had dreamed of and fought for since he set up his first real laboratory in San Francisco in 1926. In Maine, he finally had a well-equipped facility completely at his disposal, funded by the corporate fruit of his tireless labors. With just three associates working in close concert, the operation remained small and nimble, able to respond quickly to new ideas or unanticipated obstacles. More important, Phil was at last far from the prying eyes or interference of corporate masters with other priorities. Nevertheless, the company continued to benefit from the valuable projects he pioneered, including a radar jamming device the Allies used to disrupt German reconnaissance during the invasion of Normandy in June 1944.

Here at last Philo T. Farnsworth had the time and the means to explore the intriguing phenomena he had observed over the years. At last, he turned his undivided attention to the brilliant, star-like spectacle he had observed in the Multipactor and Klystron tubes.

Farnsworth concluded that there were two factors contributing to the effect. First, he determined that the currents within both types of tubes ionized residual gasses, causing electrons to leave the orbit of their nuclei and create a cloud of negatively and positively charged particles called "plasma." Second, he realized that the ionized plasma was assuming a spherical shape, held together by the electrostatic forces within the tube. That the plasma was contained within this charged sphere explained why the brilliant manifestation never touched the walls of the tube that produced it. Informally, the phenomenon became known as "The Farnsworth Effect,"[103] and Phil knew instinctively that the phenomenon opened a new world of intriguing possibilities.

If there was a central theme to Farnsworth's explorations during the 1940s, it can be found by tracing the trajectory that carried Farnsworth to the outer edges of Albert Einstein's universe. From the very beginning, Einstein's theories had shaped Phil's own ideas and manner of thinking. Indeed, television was born in part of Einstein's writings, which described the relationship between matter and energy as well as defining the quanta where light and electricity interact. Einstein was Farnsworth's constant companion throughout his career, a revered but distant mentor, the theoretician looking over the inventor's shoulder and invisibly guiding his hands. Albert Einstein was the unseen artist sculpting the shape and texture of Farnsworth's thought processes. Now the student was getting ready to surpass the teacher.

Today, the Farnsworth family speaks in guarded, almost reverent terms about the ideas that took root and shape in Farnsworth's mind during his sabbatical in Maine. He was working with "relativistic math," the sort of cosmic calculus that hints at the keys that can unlock the secrets of the universe on both a macroscopic and microscopic scale. As early as 1926, Phil had told Pem that they would one day travel together among the stars. Years later, another family member cautiously implied that Farnsworth was exploring a theoretical region where manipulating atoms revealed unexpected insights into the mysteries of traversing the perceived vastness of outer space.

Farnsworth's elder son, Philo III, was a frequent participant in dinner-table conversations in which his father would sometimes give flight to his most daring musings. According to Philo III:

> "He was into cosmological math, into cosmology, and was always cooking there. Every once in a while he'd get to a place where more ideas would occur to him and he'd start talking about it, start making some observations. He was very influenced by Einstein and all that, but these discussions often went beyond Einstein, into a realm of his own personal math, and his own personal cosmology."[104]

Clearly, mealtime conversation for the Farnsworth family was not what you would expect to hear at your typical American dinner table. The family's recollections about what came up during these discussions shed some light into the areas that were gestating in Farnsworth's imagination. Their reflections on those conversations also suggest the downside of where his ideas were taking him.

There is an inherent danger to surfing down the face of the biggest waves of modern science: as with any field of endeavor, once you reach the top, it gets pretty lonely there. Once you reach the uncharted territory where "no man has gone before," the next thing you discover is that there's nobody around you can talk to about it. The farther out Farnsworth went, and the closer he got to the cosmological Sirens that called him, the more he found himself totally on his own; the more his ideas began to escape the gravity of existing knowledge, the more he had only his own internal compass to guide him.

In spite of these explorations or perhaps, in part, because of them, Farnsworth's fluctuated dramatically throughout the war years. He never fully recovered from the breakdown of 1940 and '41 that drove him to escape Fort Wayne. Then, as the war neared its likely conclusion in the latter days of 1944, the company's board of directors appealed to Farnsworth to return to Fort Wayne to help negotiate the transition from military production to television and radio manufacturing. When Farnsworth refused the request to return, the board suspended its support of his facility in Maine and demanded that Cliff Gardner and the other employees working with him return to Indiana. Rather than return without Phil, Cliff and his co-workers turned in their resignations, and Cliff was eventually compelled to find work running the tube department of Raytheon.

Once again, Phil started selling company stock in order to maintain his research operations. But the loss of Cliff Gardner, after nearly two decades of working together in their close professional partnership and personal fellowship, was more than Phil could bear. Cliff's departure from the fold drove his brother-in-law once again down the steep slope of depression and self-medication with alcohol. Phil was bedridden again for weeks, "bent on self-destruction," in Pem's words. By the time a doctor examined him, Phil's condition had deteriorated to the point that he was diagnosed with peripheral neuritis—damage to the nerves in his legs and feet so severe that doctors predicted he would never walk again.

Only after a rigorous and protracted program of physical therapy, with Pem at his side through every painful step, was Phil able to regain his strength and overcome his doctors' prognosis. His recovery still incomplete, he suffered a hernia, and by the time he recovered from the surgery to repair the hernia, he had become dependent on pain relievers. That dependency required residence in a sanitarium—and shock treatments—before he could achieve yet another recovery.

Despite all these setbacks, despite the endless ups and downs of his physical condition, something vital was still brewing deep in the inner recesses of Phil's mind. We get a clue of just what that was after the bombs were dropped on Hiroshima and Nagasaki, when the whole world had its nose rubbed in the frightening power of nuclear energy. It's about this time, Pem recalled, that she first began to hear Phil use the word "fusion."

SEVENTEEN

It's My Baby

Farnsworth with a deluxe "Capehart" console unit, ca. 1948

"I would rather fail at something destined ultimately to succeed than succeed at something destined ultimately to fail."
—Woodrow Wilson

At the end of World War II, the Farnsworth Television and Radio Corporation found itself operating a total of seven factories throughout Indiana.[105] These acquisitions seemed prudent at the time, because the new plants contributed greatly to the company's prosperity through the war years. But the company also took on bank loans to the tune of $3 million to finance its rapid expansion. When those loans came due at the end of the war, the banks agreed to extend them for an additional year, but the company was already experiencing a precipitous decline in its earnings. Farnsworth Television and Radio—not unlike many companies that had geared up to contribute to the war effort—suddenly found itself in a mad scramble for survival.

Phil learned that the company was in trouble in the fall of 1947, from a letter from George Everson, who still served on the board of directors and still performed his familiar role as intermediary between Farnsworth and the financial interests that sustained him. George repeated the board's appeal to Phil to return to Fort Wayne, believing that the presence of its primary asset—its founder and guiding light—could help reverse the company's declining fortunes. At the very least, George implored, Phil might consider attending one of the company's quarterly board meetings in New York to get some sense of what was going on.

Phil and Pem were already giving some thought to the coming winter. In Maine, winter temperatures not only would dip into double-digits-below-zero but would often stay there for weeks at a time. Seeking a somewhat less severe climate, they rented a furnished home in Newton Center, near Boston. Arrangements were made to close up the house and lab in Maine in October, but there was one final detail that needed taking care of. After almost a decade of building on and improving the property, they had done nothing to upgrade their insurance. They had only the original $20,000 policy on the old house. Before leaving Maine, they made an appointment with their insurance agent to rewrite the policy.

Phil was seriously considering George's request to come to New York for the next board meeting and was making the necessary arrangements for the trip when a more dire circumstance demanded his immediate attention: A fire started accidentally at a lumber yard more than six miles away from the Farnsworth homestead, and it was raging out of control, consuming vast tracts of woodlands throughout the county. Phil was on the phone with George, trying to explain that he had matters somewhat more pressing on his hands than the board meeting in New York, when an operator broke into their conversation and informed Phil that the wind had shifted, and the blaze was now headed directly toward his property. They had less than twenty minutes to get out.

A Farnsworth table model television receiver from 1948

Once the flames came over the hill, the fire raced through the clearing and consumed everything in its path. The entire Farnsworth property—the carefully restored and expanded house, the elaborately equipped laboratory, and everything in between—was vaporized in a matter of minutes. The family narrowly escaped the inferno with their lives.

Returning two days later to survey the damage, Pem and Phil found the devastation complete. Where the house and laboratory had stood was rubble and ash. A massive fireplace in Phil's second floor study had crumbled when the steel beam supporting it melted, cascading the stone chimney through to the floor below. Phil's personal library of more than 200 handpicked volumes was another pile of ash. In the ruins of the lab itself, they found only "melted globs of metal and glass."[106]

The fire was equally ruinous for the family's finances: The fire had struck just two days before the scheduled appointment with their insurance agent. They kept that appointment sometime after the fire, but by then the situation was, literally, a case of trying to buy insurance after the barn had burned down. All they could collect for their enormous loss was $20,000.

Once Phil and Pem settled into their winter accommodations in Newton Center, the full measure of that loss settled on Phil, and once again his still-fragile constitution was pushed toward a too-familiar precipice. But this time Phil seemed to recognize that the fire signaled the arrival of another big change in his life, as two previous fires had. The fire that destroyed his lab in San Francisco came just before the transition that had put Jess McCargar in charge of his fortunes; the fire that destroyed the Crystal Palace in London occurred just before the transition that diminished McCargar's control. Depressed as the fire in Maine made him, Phil recognized the beginning of another transitional period, and on some level, he looked forward to its arrival. He had big things catalyzing in his mind now, and he was growing eager to see where they were going to lead him.

The next time George called and asked Phil to attend a board meeting, Phil was ready to go. Two months later, in December 1947 Phil made a visit to New York for his first board meeting in nearly seven years. At the meeting, he learned just how troubled the company had become after the war. He agreed to make a trip to Fort Wayne to get a firsthand look at the situation.

What Phil discovered when he arrived in Fort Wayne was just how difficult a time the company was having as it tried to convert its facilities from military hardware to the peacetime production of television and radio sets. While competitors like RCA, Zenith, and Philco were finally beginning to churn out television sets in substantial numbers, the purchasing agents for Farnsworth Television and Radio found themselves oddly unable to procure the necessary raw materials to keep the assembly lines humming. When Phil inquired why this was happening, Nick Nicholas laid the blame squarely on an old nemesis: David Sarnoff.

As Nicholas explained it to Phil, Sarnoff was using his personal influence, and RCA's clout over the supply chain, to persuade suppliers to make certain that orders from the Farnsworth company would always be filled last. The implication that Nicholas conveyed was that Sarnoff's vendetta was no longer directed only at Farnsworth, who had successfully defied Sarnoff's policy regarding payment of patent royalties, but at Nicholas—the former personal assistant and trusted corporate servant who had jumped ship and joined forces with RCA's mortal enemy.

Personal vendetta or not, the fact remained that in the years immediately following the war, Farnsworth Television and Radio found itself caught in a death spiral: Unable to obtain the parts it needed to keep the assembly lines running, the company could not produce or sell enough products to meet its revenue projections; unable to meet its revenue requirements, there was insufficient cash on hand to service its debt. The company was trapped between excess capacity on the one hand and onerous debt on the other. When the banks finally called the loans, the board found itself staring at a demand for $3 million in cash that it simply did not have.

E.A. "Nick" Nicholas shows George Everson a 1948 Farnsworth TV set

Farnsworth spent two weeks in Fort Wayne, and a week in January 1948 at a convention of Farnsworth dealers in Chicago, where the company's first line of television sets was introduced. There, Pem had arranged for the two of them to stay in the luxurious John Barrymore Suite at the Ambassador East Hotel. By day, they used the suite's elegant living room to entertain guests and reacquaint themselves with colleagues they hadn't seen for years. By night, they availed themselves of the sumptuous private chamber and its romantic feather-quilted bed, where they recaptured a bit of their youth;[107] nine months later another son, Kent, was born.[108]

Phil found himself energized by the flurry of activity at the dealer's convention. Talking with the dealers, hearing firsthand how commercial television was finally exploding across the landscape, provided the perfect antidote to the discouragement he was still feeling after losing everything to the fire in Maine. Orders for new inventory were brisk. The spirit he witnessed among the dealers, and the faith they conveyed not only in the company but in him personally, impressed upon Phil that he still had an obligation to the stockholders to help the company through its difficult transition. Much to Pem's surprise, he announced in Chicago that he would return to Fort Wayne later that year.

That Phil was ready to get back into the fray should have come as no surprise. There was certainly no reason to stay in Newton Center longer than the end of winter. With the arrival of spring there would be little reason to return to Maine, at least not until Mother Nature had been given a chance to rejuvenate the scarred mountains around their land—and that could take years, if not decades. Farnsworth knew his presence could help the company get back on firmer footing. But the real reason for the decision might have been something less obvious: After nearly eight years in the woods, Philo T. Farnsworth was starting to hatch some really big new ideas, and the fire in Maine meant he would need a laboratory where he could experiment with them.

After the dealers' convention, Phil and Pem returned to their temporary residence outside of Boston. From there, Phil was able to attend board meetings in New York, and monitored the situation in Fort Wayne as the death spiral that gripped the company accelerated. The management tried to stop the bleeding by slashing expenses, and it sold some of its assets to raise cash. Eventually all of the plants outside of Fort Wayne were sold. Then the company tried to arrange lump sum payments from some of its patent licensees, in particular offering RCA the use of Farnsworth's patents "in perpetuity... in consideration of the payment of $2,500,000 cash." RCA refused the offer, but did see fit to purchase one of the Farnsworth factories and made it the primary center for its television picture tube production.[109]

Despite Phil Farnsworth's re-engagement in the fortunes of the company bearing his name, a substantial portion of his mental energies continued to indulge in his cosmological explorations, especially in his extensions and interpretations of Einstein's theories of relativity. And the idea that most intrigued him was the notion of harnessing the process called fusion.

Enrico Fermi's atomic pile and the horrific bomb that it helped spawn employed a process called *fission*, which alters the atomic structure of certain heavy elements such as uranium by splitting them into other, lighter elements. The total mass of the new,

Farnsworth dealers inspect the first line of console receivers

lighter elements released by each instance of fission in a chain reaction is less than the mass of the original, heavier element. That difference in mass is released as energy in accordance to Einstein's formula $E = MC^2$. Translated, the famous formula says that the amount of energy (E) released when matter is decomposed in this manner is equal to the lost mass (M) multiplied by the largest quantity known to man—the speed of light (C) squared.

In other words, a very little bit of material can release a *lot* of energy. In the case of Fermi's reactor, this process was controlled so that this enormous release of energy would be gradual, sustained over a period of time. Converted to a bomb, the energy is released in a single blinding flash, producing enough explosive force to level entire cities.

Another way to release this tremendous energy is with a process called *fusion,* because it involves the combining of light atoms such as hydrogen into a heavier element such as helium. It is no exaggeration to suggest that fusion is the Creator's own way of releasing the energy that is locked up in the nuclei of atoms. Our sun and all the stars in the heavens are natural fusion reactors—great, celestial globes of hydrogen gas, gravitationally gathered around an atomic furnace that forces the light hydrogen nuclei to combine to form heavier helium nuclei. As with fission, the mass of the resulting helium atoms is less than the mass of the original hydrogen atoms, and the difference is given off as energy according to Einstein's formula.

There are big, important differences between fission and fusion. Fission, which does not exist anywhere in nature, produces radioactive by-products that can take hundreds of thousands of years to decompose. The only by-product of fusion is helium—a harmless, inert gas. Equally important, the hydrogen fuel that can power a fusion reactor can be easily distilled from water. By the mid-1940s, scientists in the United States and elsewhere had calculated that if fusion could be harnessed, there is enough water in the planet's oceans to furnish our energy needs for hundreds of millions of years—long after our own sun has burned off into a cosmic meatball.

Simply put, fusion offers the potential of a clean, safe, and virtually limitless source of power that could supplant all known forms of energy production. Fossil fuels, solar power, and hydroelectricity would all be obsolete in a fusion-powered future.

But fusion was proving to be much more difficult to control than fission. Controlling fusion requires nothing less than bottling a star. The task presents a riddle of cosmic proportions: How can you bottle a star without either melting the bottle or snuffing out the star? This is precisely the riddle that attracted the mind of Philo T. Farnsworth in the mid-1940s.

Of all the expansive ideas swimming around in his mind, this was the one with the most dramatic and compelling possibilities. And so, in much the same way that the idea of sending pictures through the air had captivated him as a teenager in the 1920s, Philo T. Farnsworth in the late 1940s was thinking about how he could use everything he had learned over the intervening twenty years to harness fusion.

Still, as potentially big an idea as fusion was, it was only part of a much larger cosmos that Farnsworth was wandering in. Though duly impressed with the benefits that fusion might offer for mankind, he was also seeing how its realization might lend credence to some of the more esoteric ideas that were forming in his head. As intrigued as he was at the prospect of supplying the world with a boundless source of industrial energy, he was equally obsessed with the notion that such a discovery would reveal marvels that he was beginning to believe about the inner and outer workings of Einstein's universe. He was also certain that fusion would provide an efficient means for traveling in outer space.

This was an immense, convoluted corridor that was carving its way through the gray matter of Philo Farnsworth in the mid-forties. This time, however, there was no Justin Tolman he could turn to, no benevolent tutor he could spend afternoons with, drawing sketches on chalkboards and doodling diagrams in sketch pads. Absent a reasonable sounding board, he knew that at least one small corner of his brain had to keep a watchful eye on his sanity.

All that changed one afternoon in the summer of 1948, when Phil and Pem were in New York for another board meeting. They visited the home of Frank Reiber, a friend from their San Francisco days. Phil was surprised to find the Reibers' apartment adorned by a portrait of Albert Einstein that had been painted by Reiber's mother—quite possibly the only time Einstein had ever agreed to sit still long enough to have his picture painted.

Admiring the painting, Farnsworth shared a secret desire with his friend. "You know, I'd love to talk to Einstein some day about some of these ideas I've been thinking about. But I really need to get some of my math together before I can do that."

Reiber had a better idea. "Why wait?" he said. "Let me see if I can get Einstein on the phone for you."

Reiber disappeared into the bedroom, and reappeared a few minutes later saying, "I've got Einstein on the phone, and he'd be delighted to talk to you."

Nearly an hour later Phil emerged from the bedroom, "his face aglow from the excitement of finding someone who knew what he was talking about."[110] We will never know exactly what these two eminent scientists—the theorist and the inventor—discussed during that hour. We can only surmise that they walked

together along a path that very few others can even imagine, let
alone discuss intelligently.

At the time of this conversation Albert Einstein—self-exiled
from Nazi Germany and a lifelong pacifist—had all but ceased his
further studies, so disturbed was he after his theories were applied
to flatten Hiroshima and Nagasaki. But when Farnsworth told him
the direction his thinking was taking him, Einstein told Phil that he,
too, had been thinking along similar lines. When he told Einstein
that he had some rudimentary ideas about how to harness the
power of fusion, Einstein is reported to have replied, "You must
continue this work. This is the good part of my theories."

When Farnsworth left Frank Reiber's apartment that day, he
was greatly encouraged that his own ideas were sound, and more
determined than ever to follow the path he was on to some kind
of fruition. He had also promised Einstein that he would publish
his math as soon as he could get it a bit more formalized and
suitable for publication.

First he had to save his company.

When he returned to Fort Wayne that August, Farnsworth found
the company in even more dire straits than it had been nine
months earlier. Management was continuing to slash expenses,
including a 35 percent reduction in its advertising expenditures,
and continuing to liquidate assets.[111] Even the Fort Wayne radio
station was sold, ending Phil's long-held dream of creating a
Farnsworth broadcasting network. Still, the company continued
losing money on its operations and coming up short of the cash it
needed to pay off its creditors. Suppliers were beginning to with-
hold credit, demanding cash on delivery of raw materials, making
it even harder for the company to keep its assembly lines rolling.

Phil was welcomed to Fort Wayne with the sort of reception
worthy of a returning hero, people shaking his hand as he navi-
gated the hallways, appealing for snippets of his time. He went
right to work, immediately breathing new life into the company
that bore his name. Calling on friends and contacts throughout
the industry, Farnsworth personally arranged for dozens of new

research contracts that restored vitality to the Fort Wayne laboratory and infused management with renewed vitality.

This activity was a tonic to Phil personally, as well, restoring what was still missing of his own energy and strength. But even the added cash flow from Phil's projects was not enough to reverse the tide. The bank loans still loomed and the company still lacked sufficient cash to pay them off.

Something else Phil learned was that buyout offers were on the table from RCA, General Electric, and a new player in the arena, a huge conglomerate called International Telephone and Telegraph (ITT). The board was seriously considering these offers, as well as the possibility of declaring bankruptcy, when Nick Nicholas and the board came up with one more strategy that might succeed in forestalling all of those possibilities.

In January 1949, Wall Street underwriters for the Farnsworth Television and Radio Corporation filed a registration statement with the Securities and Exchange Commission for the sale of a secondary issue of new stock in the company. The offering of new stock would dilute the value of those shares already outstanding, but it was expected to raise enough fresh capital to meet the company's obligations and allow it to stay in business— or so Nicholas thought. Unfortunately, just as the underwriters were ready to go public with the new shares, the SEC discovered discrepancies in the company's financial statements, and declared that trading in all FTR shares be suspended for one hour on January 14 while their inspectors looked over the documentation.

The discrepancies were explained, and trading in the company's stock resumed, but the suspension caused irreparable damage to the company's credibility. The secondary offering was withdrawn, forcing the board to look elsewhere for a resolution of its financial difficulties. Only this time, the options were much more limited. The acquisition offers from RCA and General Electric had been withdrawn, leaving the board with only one suitor willing to take the company off their hands. ITT tendered an offer to exchange shares of its stock for all outstanding Farnsworth shares for what amounted to a total valuation of approximately

$1.4 million—a fraction of the company's real value. Dreary as the offer from ITT was, it still looked better than bankruptcy and liquidation, and the board prepared to accept the ITT offer.

That action led to the company's principal asset—its titular founder, Philo T. Farnsworth, suddenly finding himself pressed into the unpleasant service of standing as the company's most visible representative as it entered the final phase of the death spiral. When an angry contingent of stockholders organized in opposition to the proposed sale to ITT, Nick Nicholas prevailed upon Phil to write a letter to the stockholders, recommending they vote to accept the ITT offer. The letter, distributed just prior to a meeting of stockholders called to discuss the ITT acquisition, implored the stockholders to vote their shares, as Phil was prepared to do with his. He sympathized readily with the plight of the other stockholders because the deal meant a virtual wipeout of the value that was left of his own stake in the company.[112]

Just as the meeting of the stockholders was about to convene in Fort Wayne, Nick Nicholas and his deputy, chief counsel Ed Martin, inexplicably found themselves called away on "urgent business" elsewhere. The disgruntled stockholders assembled to find only Phil Farnsworth and another lawyer left to face their anger. When Pem later asked Phil why he was willing to submit himself to such humiliation, Phil could only say, "It's my baby, so I guess it's my job to bury it."[113]

In the last months of the 1940s, in the era of unbridled prosperity that followed World War II, the newest feature on the American landscape was the television antenna, sprouting like saplings on rooftops all over the country. Uncle Milty, Jack Benny, Sid Caesar, and the other stars of television's true dawn, together with the be-the-first-on-your-block monochromatic small screen, all became fixtures in American households.

But in Fort Wayne, Indiana, the company formed in the name of the man who breathed life into all these living room dreams was sold to a multinational conglomerate, and the Farnsworth Television and Radio Corporation vanished from the New York Stock Exchange.

EIGHTEEN

Stars in a Jar

Farnsworth prepares the first Fusor for testing in 1960

"The release of atom power has changed everything
except our way of thinking... the solution to this
problem lies in the heart of mankind. If only I had
known, I should have become a watchmaker."
—Albert Einstein

In the weeks and months immediately following its acquisition
of the Farnsworth Television and Radio Corporation, the management of ITT expressed its intent to continue what was now
the Farnsworth "division" with its own identity, with ITT serving
merely as a distant parent company. But within months the name
of the unit was changed to Capehart-Farnsworth Company, drop-
ping the word "television" from the nomenclature in an exercise
that portended things to come.

At the time of the Farnsworth acquisition, ITT was beginning
to undergo its own corporate metamorphosis, beginning with the
elevation of its elderly founding CEO, Sosthenes Behn, from the

operational position of president to the more ceremonial role of board chairman. After Behn's ascension—retirement, really—the company went through numerous top-level management changes. With each change in the corporate executive suite, a new president would be assigned to the Farnsworth division, its name was changed, and the unit's mission was redefined. The only constant throughout the years was the presence of Philo T. Farnsworth himself, prompting one co-worker to observe that "ITT presidents come and go but Farnsworth goes on forever."[114]

What did not go on forever was the company's manufacture of television sets. By the time ITT took over, the Farnsworth line of receivers had been steadily losing market share to its competitors, especially RCA and Zenith. So ITT divided its new subsidiary into two smaller units, leaving the Capehart-Farnsworth Division in the manufacturing business and creating Farnsworth Electronics as a research-only operation focused on the procurement of government-sponsored, Cold War defense contracts. Eventually, the manufacturing operation was sold off completely, marking the formal, corporate end, once and for all, of Farnsworth's involvement with television.

Farnsworth was largely unfazed by all the corporate machinations going on around him. He was grateful to be relieved of the responsibilities that went with being part of a company that bore his name. He seemed to relish the work made possible by his ITT superiors, as it kept his hands busy and one part of his mind occupied, while the other, visionary part was allowed its slow and deliberate focus on the mathematical and cosmological perplexities that continued to absorb his deeper imagination. In his lab, he conducted research with various kinds of vacuum tubes, experimenting with "photo-emissions, image converters, photomultipliers, advanced Image Dissectors, and image storage devices."[115]

But at home, he spent long solitary evenings in his study, running complex calculations through a MonroeMatic four-function motorized calculator, refining the math that he had promised Einstein he would publish. The MonroeMatic was the personal computer of its day, an ungainly machine that worked like a slot machine, requiring Phil to pull a lever every time he was ready to test an equation. In his head he was manipulating the celestial forces within atomic nuclei, and all he had at his disposal to confirm his rarefied theories was a one-armed-bandit of an adding machine.

When young Philo T. Farnsworth started thinking about television in the 1920s, he devoured all the information he could find on the subject. Starting with Einstein's paper on the Photoelectric Effect, he taught himself about electrons—how they could be bounced around by magnets and how a cathode ray tube could convert them into light. These are the ideas that converged in his mind, finally synthesizing in the conception of the Image Dissector, which eventually produced the system of television still in use around the world at the dawn of the 21st century.

Similarly, with thirty years of firsthand experience in putting electrons through hoops of his own design, Farnsworth was assembling the information he would need to solve the riddle of nuclear fusion. He had taught himself all he could about the properties of protons, electrons, and neutrons, and how they could be manipulated. That very specialized knowledge began to germinate in the early 1950s in the fertile soil of his gifted mind. He had already performed reams of mathematical calculations to confirm that he was on the right track. And he began to see how The Farnsworth Effect—the shimmering blue star in his Multipactor and Klystron tubes—was the key that would unlock the new kingdom

In the summer of 1953, Farnsworth was invited by the Utah Broadcasters Association to attend a banquet in his honor and accept the recognition of peers from his native state for his contributions to science and communications. This honor he accepted, on the condition that Justin Tolman and other individuals who were important to his formative years be present to share the honor with him. With all arrangements for a memorable event in place, the entire Farnsworth family piled into a 1949 Cadillac and headed across the Great Plains for a week in the mountains of Utah.

Philo III recalled the trip vividly: Rolling over the hot, endless plains in an "unrefrigerated car," he shared the backseat with his new wife, Ruth, his younger brother Skee, and a fifth of white wine, which they shared "in the clutch of our own ennui, hoping the day will end as soon as possible." Pem was driving, with four-year-old Kent asleep with his head in her lap. Phil was slumped in the front seat, his head down, his fedora pulled down over his eyes.

All of a sudden, "Dad practically jumped out of his seat in one fluid movement and punched his fist forward, saying 'I've got it.' It was very uncharacteristic of him to grab you like that and say 'hey, I've got it' to a car full of people. And I knew instantly . . . my brother Skee and I had heard a lot of Dad's talk . . . we looked at each other and knew instantly that he'd had a large conception."

As the concept of electronic television had arrived in a potato field in the summer of 1921, a practical approach to fusion energy arrived in a '49 Cadillac on a Great Plains highway, somewhere between Indiana and Utah, in the summer of 1953. Farnsworth's math, his conversation with Einstein, his years of experience, his observations in the Multipactor—and most of all "the daring of this boy's mind"—suddenly converged to deliver the concept for a device and a process that could unleash the power of the atom cleanly and safely, with a fuel source as abundant as water. As his son concluded, in that moment in the car somewhere between Indiana and Utah, Philo T. Farnsworth had discovered "the answer to the riddle of the sphinx."

There was not a great deal more said that day, although everybody in the car knew they had been present for an important moment. The Cadillac filled with Farnsworths continued on its way across the plains, on toward the mountains and valleys of Utah, with a precious bundle of new cargo riding in the front seat.

In Salt Lake City, Phil was warmly received back into the cradle of his early years. Before the banquet, Phil was reunited with Justin Tolman, the high school chemistry teacher who had played such an instrumental role not only in his conception of television,

but in the defense of his patents. Also present was another of his revered teachers,
Frances Critchlow, as well as
teachers from his BYU days
and elders from the Mormon
Church. The whole affair gave
Phil and Pem a heartwarming
return to the people and values they cherished most, and
they were truly grateful for
the generous outpouring of
fellowship.

After the dinner, the president of the Utah Broadcasters
Association presented Phil

Justin Tolman with Farnsworth in 1953

with a large, ornately inscribed plaque commemorating his
achievements and contributions. As he rose to accept the honor,
his heart was filled with gratitude, but his soul was burning with
another kind of excitement. In the wake of that moment in the
car two days earlier, the prospect of harnessing fusion was as real
to him as the concept of electronic television had been when he
was a teenager. Thirty-some years later, fusion was within his
grasp, and with it, a new future of astounding possibilities.

Standing at the podium to share a few remarks, he looked
out over the tables filled with family and lifelong friends. He
tried to stay focused on the moment, but as he spoke his
thoughts faded in and out, from past to present to future. And
he was possessed by another singular insight. As clearly as he
had foreseen television and the changes it implied when he was
a teenager, he now foresaw a world powered by fusion energy
with equal clarity and resolution. As he thanked his audience for
the honor they were bestowing upon him and for the recognition that served to bring them all together, he was consumed
with an entirely new vision of an even more fabulous future that
only he knew awaited them.

When he returned to Fort Wayne after a short vacation in Utah, Farnsworth did not rush right back to the lab to build a fusion device, any more than he'd rushed back to his attic loft in 1921 to build an Image Dissector. His moment of conception had unlocked the door to the kingdom, but there was still a serpentine path to navigate before he could reach his Holy Grail. He still had a great deal of work to do before he could seriously propose to his new bosses that they invest in something as radical as a nuclear fusion reactor.

With his flash of insight as a beacon guiding him into the diaphanous distance, Farnsworth spent the next six years contemplating and refining his inspiration, grinding figures through his mechanical calculator, confirming every aspect of the concept on paper before he would even attempt to actually build it.

If you've ever sliced open a golf ball, you have some idea what Farnsworth was formulating: Where the golf ball winds densely packed layers of rubber bands around its cork core, Farnsworth figured out how to organize the hot, agitated atomic soup called plasma into densely packed, alternating layers of electrons and protons. He believed the plasma would take shape within two concentrically arranged, softball-sized stainless steel spheres, one called an anode and the other a cathode. Through openings in the inner sphere, he proposed to inject the nuclear fuel, a common form of hydrogen called deuterium. As the deuterium nuclei negotiated their way toward the center of the plasma, they would become packed so tightly they would "pop" together, like tiny soap bubbles, to form helium nuclei. The helium nuclei, having less mass than the nuclei of their hydrogen parents, would release the difference as energy, according to Einstein's famous formula.

Farnsworth knew instinctively that this reaction, this tiny, synthetic star, would neither melt the walls of its container nor be cooled by them. The plasma would reach celestial temperatures at its reactive core where the fusion takes place, but the reaction would suspend itself within its own isolated space, precisely as he observed of the star-like blue glow in his Multipactor and Klystron tubes.

To prove his theories, Farnsworth spent hundreds of hours experimenting with the equations of Simeon Poisson, an 18th century French mathematician whose theories explored arcane matters of probability. He applied Poisson's equations to his charged spheres and, when he was satisfied that his concept would work, decided to call his plasma ball a "Poissor" in his predecessor's honor. Because the process he envisioned would employ the particles' own inertia to cause them to fuse within the electrostatic fields of the Poissor, he called the entire approach "Inertial Electrostatic Confinement," or "IEC" for short. Finally, in another demonstration of his fondness for simple and descriptive nomenclature, he decided to call his new creation the "Fusor."

As was typical of Farnsworth's previous inventions, the solutions within the Fusor were unique, simple, and elegant. But as the design of the Fusor became more evident to him, the inventor found himself wrestling with much more than the physics and math of his conception. He wrestled with the metaphysics of it, the cultural implications, and the economic ramifications.

On some gut level, the whole proposition scared him. At the same time he was reinforcing his math and physics, he went through a period of moral uncertainty unlike anything he had experienced in his life.

In soul-searching conversations with Pem, he "honestly questioned whether he had the right to turn this kind of a powerful thing loose on an ecology that wasn't ready for it." He worried about how fusion might be used to diabolical ends. He feared that fusion had the potential to create frighteningly dangerous new weapons, even musing to Pem on more than one occasion that "a fusion powered laser could bore a hole right through the moon."[116]

Who was he, Farnsworth demanded of himself, a mere mortal, to create an artificial sun? He knew well the ancient Greek mythology: Was he on the path of a modern-day Prometheus, challenging the gods, stealing their fire? If he succeeded, would the gods respond as Zeus had, by opening Pandora's box and unleashing great horrors upon the world?

Then he recalled his conversation with Einstein: "This is the good part of my theories," Einstein had said. "You must continue this work." Suddenly he knew what the old professor must have meant. Pandora's box was already opened. The horrors of nuclear power had already been unleashed—by Fermi and Oppenheimer and Teller, at Alamagordo and Eniwetok, and Hiroshima and Nagasaki.[117] What worse could come of the nuclear genie than the atomic bomb, the hydrogen bomb, or lethal fission reactors?

We have already paid the price, Farnsworth concluded. He reasoned that controlled fusion is the only nuclear equation that maximizes the upside and minimizes the downside. True, a benevolent society might have arrived at controlled fusion first, rather than going straight to the ultimate weapon. But there was nothing he could do about that. Now it was time to put the genie in a bottle, to harvest the benefits mankind had already paid for.

All this festering over the possible impact of his new discovery came with its own price for Phil Farnsworth. Early one morning in March of 1955, he awoke with a searing pain in his gut. Pem called the family doctor, who came to the house and diagnosed a bleeding ulcer. While an ambulance rushed Phil to the hospital, Pem delayed following him only long enough to make arrangements for the care of little Kent. By the time she reached the hospital, a surgeon had already taken aggressive action, performing an operation that removed nearly two-thirds of Phil's stomach. Eventually he recovered from the surgery, but forever found it difficult to digest his food. The ill-advised procedure was another setback for Phil's always-fragile health.

While recovering, Farnsworth was saddened to learn of the death of Albert Einstein in April 1955. He had so hoped to confer with Einstein one more time, to go over the math with him before he published it. But he had not had the time or opportunity to "formalize" his mathematical ideas so he could discuss them with Einstein again. Now it was too late. Without a chance to talk to Einstein again, Farnsworth was uncertain if he'd be able to fulfill his promise. And he had lost the only other man on Earth who might be able to travel alongside him on his voyage through the cosmos.

By 1959, after six years of burning the mathematical midnight oil, Farnsworth was prepared to test his ideas on a workbench. He was ready to build his first Fusor. He presented his ideas to his superiors at ITT. Perhaps not surprisingly, his bosses did not share his enthusiasm for this exotic new line of research.

Although the ITT executives might have grasped the importance of what Farnsworth proposed, they never quite grasped the subtlety of his approach. Rather than make a decision themselves, they deferred to the supposedly higher authority of the Atomic Energy Commission. But the AEC was unfavorably predisposed toward the notion of electrostatic confinement, and Farnsworth's initial proposal did little to dissuade the commission of its bias.[118]

The authorities at the AEC thought they had already figured out how to do fusion. At the time Farnsworth proposed to build his first Fusor, the generally accepted approach to controlling fusion—the approach the AEC understood, endorsed, and funded—employed the brute force of giant magnets and coils to compress and heat the fusion fuel; in effect, using massive external forces to create and control the artificial star. The Fusor, on the other hand, was designed to work from within, to exploit the properties of the particles themselves.

Farnsworth predicted that once the "Poissor" became fully formed, the resulting star-like clouds of nuclear fuel could be confined without the need for any kind of external magnets. But Farnsworth's ideas were a hard sell. Despite the simple elegance of the Fusor, the AEC preferred to bestow its growing fusion research budget on monolithic machines, particularly a device of Russian origin called the "tokamak."

Compared to the tokamak, the amount of money that Farnsworth needed to build his Fusor was very conservative. What cost tens of millions of dollars to develop around the concept of magnetic confinement might cost hundreds of thousands with Farnsworth's IEC approach.

As he tried in vain to secure the funding he needed to test his theories, Farnsworth began to realize just how far he was from the mainstream, and how entrenched that mainstream was when it came to considering new ideas. What had once been an unguarded frontier and an open-minded fraternity had became classified during the war, and was now the exclusive province of those physicists who had pioneered nuclear research in early 1940s and their various institutional descendants and heirs. A rigid orthodoxy had evolved around those institutions and its adherents, and outsiders with unorthodox ideas like Philo T. Farnsworth often found themselves unable to obtain even the funds their experiments required, no matter how modest.

By this time, Farnsworth could care less whether the entrenched scientific community or his superiors at ITT were interested in what he wanted to do. He was determined to proceed, whatever the personal cost, even if it meant setting up a fusion lab in his home. The device had to be built, if only to prove to himself that the scientific limb he'd gone off on was a solid one. He needed to build the Fusor to determine once and for all if the ideas he had talked to Einstein about were valid.

Since their personal finances were seriously depleted in the wake of the Maine disaster and the ITT acquisition, Phil and Pem decided to take a second mortgage against their house and a loan against his life insurance. Thus funded, he started bringing copper pipes and tanks into his house and setting up a lab in a spare bedroom. He also hired Gene Meeks, a clever young electrical engineer from the ITT plant, to come in two or three evenings each week to lend a hand. They did not get very far before officials at ITT got wind of his activities and reconsidered their disinterest.

Farnsworth had at least one friend on the ITT board. Frederick R. "Fritz" Furth was a retired navy admiral whom ITT had invited to join its board in a not very thinly veiled attempt to win military contracts. One unexpected fringe benefit of the arrangement for Furth was the relationship he formed with Philo Farnsworth, for whom he had great respect and admiration. Fritz became Phil's personal mentor and advocate within the ITT corporate structure, and was personally intrigued with Phil's fusion ideas. He knew Phil well enough to know that he didn't often go down blind alleys. Fritz decided that if Farnsworth sincerely believed his ideas would work, he should be given a chance to pursue them.

Fortunately for Fritz and Farnsworth, later that year Harold S. Geneen became the president of ITT. Geneen was cut from a very different cloth from that of previous ITT chief executives, and he was similarly intrigued by what Farnsworth said he could do, though a bit reluctant in light of his larger corporate responsibilities. Nevertheless, Geneen invited Farnsworth to present his proposal for controlling nuclear fusion to the ITT board in New York. Then, despite the misgivings of the AEC and the board's own reluctance to engage in a pursuit so far outside the scope of ITT's corporate mission, Fritz Furth managed to persuade Geneen and the board to provide Farnsworth's fusion ideas with a nominal level of funding. Apparently, the board figured that if Farnsworth was that determined, perhaps they had better lend some corporate support—if only to keep all his ideas under their roof.

Almost immediately, the requirements of Farnsworth's research and the priorities of ITT's corporate interests began to diverge. Farnsworth felt the research needed its own facilities, preferably in an isolated area that would permit experiments at the very high energy levels he believed the Poissor would require to fully form. While the process of fusion is inherently safe, in the first stages of research Farnsworth knew he would have some radiation to contend with, and he wanted the facility adequately shielded to minimize those risks. During a working vacation to California, he found an ideal location, in the high desert near Twenty Nine Palms, and he recommended that the company establish a fusion lab there. ITT was not prepared at that early stage to bear that much expense, and instead instructed Farnsworth to set up shop in the basement of the existing facilities on Pontiac Street in Fort Wayne.

Given the highly sensitive nature of the work, Farnsworth felt it best to keep the new lab team small. With Gene Meeks already in place as Farnsworth's special assistant, ITT added George Bain, another electrical engineer as resident project chief and budget supervisor. By the end of 1959, work began on the construction of the first actual Fusor—the device Farnsworth had first conceived six years earlier, based on a phenomenon he had observed twenty years before in the Multipactor.

By October 1960, the very first Fusor was assembled, the fusion chamber itself situated under a clear glass bell jar that allowed visual access to the reaction chamber. When the apparatus was connected to a vacuum pump, power supply, and fuel source, it was ready for testing. The very first time power was applied, evidence of a plasma formation was witnessed. The next day, deuterium gas was introduced into the chamber, causing a Geiger counter placed adjacent to the bell jar to click rapidly and loudly. However, it was not possible with this rudimentary assembly to tell just what the Geiger counter was measuring.

What Farnsworth and his team were hoping to find were neutrons, which are the evidence of fusion. Deuterium is an isotope of hydrogen, meaning its nucleus has an extra neutron in addition to its single proton. If the proton of one deuterium nucleus did indeed fuse with the proton of another deuterium nucleus, one of the neutrons would be released, discharging itself from the new nucleus with the energy that was the object of the experiment. The more neutrons Farnsworth counted with his Geiger counter, the more fusion the Fusor was producing. Unfortunately, a Geiger counter is not a very efficient means of counting neutrons; it also registers x-rays, gamma rays, and other forms of radiation.

The following Monday—October 9, 1960—a shield was procured to filter out whatever x-ray and gamma radiation the Fusor generated. Power was again applied, deuterium gas was again admitted to the chamber, and the Geiger counter again began to register. This time, with the shielding in place, there was no doubt what was being measured. The rapid clicks of the Geiger counter could only be counting neutrons. The counts were low, but the conclusion was unmistakable: the Fusor was producing fusion.

As George Everson had written to Les Gorrell thirty years earlier: "The damn thing works." Philo T. Farnsworth

The Fusor lab on Pontiac Street in Fort Wayne, Indiana, ca. 1962

had done it again: synthesized his learning, fortified it with his own observations, added a sprinkle of his own inspired imagination, conceived a device, and walked into a laboratory to produce precisely the intended result. There was indeed a miniature, artificial star forming under the bell jar housing the Fusor—Farnsworth's first "star in a jar."

This early success was encouraging, but by no means conclusive. These first tests were conducted at relatively low power levels and lasted at best a matter of seconds—hardly enough to provide a viable source of power—any more than Farnsworth's first successful television experiments could tune in "I Love Lucy." But these experiments in the fall of 1960 proved that the Fusor could produce fusion, just as Farnsworth knew it would the moment he bolted out of his car seat with the idea.

There were still critical questions to be answered. Principal among them: Could the Fusor produce enough fusion so that the amount of energy coming out of the reaction would be greater than what went into it? To be a viable source of power, the Fusor would have to produce more energy than the sum of the voltages and currents that went into the chamber—a goal known as "breakeven." And, if such a breakeven level could be achieved, could the Fusor sustain the reaction indefinitely?

Achieving a sustained, better-than-breakeven reaction is the goal of all fusion research and represents a turning point in human history, analogous to the discovery of fire millions of years ago. It is often called "lighting the fusion torch."

Farnsworth was keenly aware of the impact that harnessing fusion would have on civilization and felt no small obligation to do whatever he could to prepare the world for the tumultuous changes that he anticipated.

We get an early taste of how significant fusion was to Farnsworth in 1957—two years before he'd actually built a Fusor—when he appeared on the TV quiz show *I've Got a Secret*. His secret: "I invented electronic television—in 1922 when I was 14 years old."[119] After the celebrity panelists failed to correctly identify the mysterious "Dr. X," Farnsworth talked a little about what he was working on at the time, suggesting only that "in the future we'll heat our homes, run our cars...all with nuclear fusion." After that startling forecast, he was handed a carton of Winstons, eighty dollars cash, and Garry Moore's words of eternal gratitude: "I'd be out of work if it weren't for you." Ironically, this was the only time the man who invented television appeared on it. But the greater significance of that event lies in the way it hints at the larger issues he was already dealing with at the time.

During the late 1950s and early 1960s, Farnsworth overcame his usual reluctance to appear in public; fusion was so real for him, he felt obligated to do whatever he could to prepare the world for its imminent arrival.

In these public appearances, he would begin by saying that the achievement of controlled fusion was more than a step in evolution, more than a historical flex point like the Industrial Revolution. Controlling fusion, Farnsworth believed, represented a quantum leap in human knowledge and understanding, the point of demarcation between two epochs of civilization. He suggested that we are now living in the "low energy" epoch, characterized by industries powered by fossil fuels consumed in a variety of chemical processes akin to fire—the same energy source that served the caveman. But once fusion was controlled on Earth, it would usher in Farnsworth's predicted dawn of a new "high energy" era, in which all energy would be derived

Farnsworth on *I've Got a Secret* in July 1957

from the same force that he believed God himself had settled on when He created the Universe.

Farnsworth spoke of "the enormous energy that would be at the disposal of every man, woman, and child on the face of the Earth." He predicted that great cities would be powered for pennies, with no pollution or nuclear waste; that every home would be equipped with its own fusion generator, much as homes today are equipped with a furnace; and that power lines would become a thing of the past. Fusion engines would replace internal combustion, making every type of travel cheap and clean. Barren deserts would be reclaimed to produce immense quantities of food. And he always spoke of how much fun it was going to be.[120]

In short, Farnsworth believed that fusion promised a level of human abundance previously unimaginable. Fusion, he predicted, would forever divorce the cost of fuel from the cost of power. Together with Fritz Furth, Farnsworth calculated that fusion could produce enough electricity to run a city the size of New York for an entire month for about a nickel.

As dramatic as these changes would be on the domestic front, Phil saved his fondest forecasts for the impact that fusion would have on space travel. He believed that fusion would alter the basic relationship that hinders current space travel, the weight ratio between launch vehicle and payload. He used the analogy of a pineapple and a pea: Today, what little space travel we do is conducted with payloads the size of a pea that are lifted into Earth orbit by launch vehicles the size of a pineapple. The reason for this inefficiency is because so much fuel has to be consumed in the initial stages of the flight just to get the rest of the fuel off the launchpad. Farnsworth predicted the reversal of these ratios, with small, fusion-powered rockets gently lifting enormous payloads into orbit. He predicted that once in orbit, fusion-powered spacecraft could make it to Mars on as much nuclear fuel as could be stored in a tank the size of a fountain pen

When his thoughts ventured into realm of interstellar travel, Farnsworth hinted at the truly daring cosmology behind much of his fusion work, and the math that he and Einstein had discussed in 1948. He dared to question our whole concept of distance as it relates to travel through outer space, asking aloud on many occasions, "Why do we assume that [we] have to exert so much energy to cross something which is actually nothing?"

Farnsworth believed that fusion would give mankind sufficient power to colonize other planets, if that was our choice, or to gather natural resources from the Moon and asteroids. He expected fusion to put our solar system within reach the way gasoline-powered automobiles put remote locations here on Earth within reach. He believed that space flight would become as commonplace as transoceanic jet flight is today. And he was drawing his own plans for a "ram jet" fusion star-drive that would hurtle through space at speeds approaching the speed of light, collecting all the fuel it needed from the random particles scattered through the vastness of space.

Before any of these glorious flights of Farnsworth's fancy could be realized, he would first have to refine his Fusor past the point of breakeven. Just as he had single-handedly refined his first television invention to the point that it became the mass medium of choice for millions of people around the world, he now faced the task of delivering a viable fusion device.

NINETEEN

That's All I Need to See

The Mark III Fusor ready for testing in 1965

"No real progress is made unless real leaps
of faith are made."
—John V.C. Nye

That the Fusor worked—that it produced fusion—was indis-
putable from that day in October 1960 when deuterium fuel
was first admitted to the reactor chamber. Whether the Fusor was
ever effectively reconfigured to produce a self-sustaining reaction
remains a mystery.

Based on this initial success, Farnsworth again proposed to
the ITT management that the entire fusion operation be moved to
the type of facility he felt he needed to push the envelope of his
experiments. He knew that in order for the Poissor—the plasma
ball at the heart of the reaction—to achieve its greatest efficiency,
it would have to be operated at power levels beyond what was

safe to work with in the basement of the ITT facility in Fort Wayne. Again, ITT refused his request, leaving Farnsworth in a quandary: Though he had succeeded in producing substantially more fusion than any other experiments anywhere in the world, the ITT people were not satisfied. They wanted to see more substantial results. To produce more fusion, Farnsworth needed better facilities. But ITT would not give him better facilities until he demonstrated that he could produce more fusion.

But in a conciliatory gesture, ITT relocated the fusion lab to new quarters at the rear of the first floor of the Pontiac Street plant. This was still not the kind of environment that Farnsworth felt he needed, but it was an improvement, providing an exhaust hood, residual gas analyzers, and a power supply capable of 100kv (kilovolts). The entire operation occupied one relatively small room. To shield the workers, the Fusor itself was eventually placed on a hydraulically operated elevator that could lower it into a fourteen-foot-deep hole that was dug through the floor of the basement. Farnsworth had wanted to operate his Fusor in an entirely separate location; ITT put it in a pit.

The Fusor underwent at least two major redesigns, called "Marks." The first October 1960 model was dubbed the "Mark I—mod. 0" (Mark one, modification zero). Each minor modification was given a new number. Complete redesigns were given a new Roman numeral. In 1961, the Fusor Mark I went through a few variations, and in 1962 was replaced by the Mark II.

With each new redesign, the neutron counts—the measure of just how much fusion the Fusor was producing—increased steadily, if gradually. Neutron counts are reported in exponential increments, as in 10^5 or 10^6; Farnsworth's first important goal was a neutron count of 10^9—that is a one followed by nine zeroes, or one billion neutrons per second. The Fusor approached this level in steady increments with each redesign and modification.

During 1962 and 1963, strange and startling things began to occur in the course of Farnsworth's continued fusion experiments—events of an epic nature suggesting that there was much more going on inside the Fusor than mere neutrons could account for. These unexplained incidents suggest that the Fusor was much closer to the goal of self-sustaining, breakeven fusion than anybody present realized or the written record reveals.

One such event is described in *Distant Vision*, Pem Farnsworth's biography of her husband:

> ...as Phil put the equipment through its paces, he increased the power input beyond levels of previous tests. The degree of nervous tension in the lab was already high, as it always was when the Fusor was being tested. Phil sat at the controls, slowly adding power...75 kilovolts...80 kilovolts ...then 90 and 95. Not a word was spoken, and all eyes turned toward Phil as the power went past 100 kilovolts. Suddenly, there was a terrific power surge, a loud crack like a high-powered rifle, and a lightning-like electrical discharge in the pit. Fearing the worst, all hands abandoned their posts and bolted for the door—except Phil, who sat calmly watching his gauges, quite certain that everything was under control.[121]

What is most noteworthy about this particular event is the reaction of the various witnesses. While Farnsworth remained comfortably at the controls, the rest of his staff beat a hasty retreat from the room. That's just one reason why events of this nature came to be known as "runaway." Farnsworth's own composure throughout the episode suggests that the experiment was under control and *not* running away. To the contrary, the discharge suggests that the experiment had reached an important threshold, something akin to an airplane breaking the sound barrier—precisely the sort of outcome that would warrant further observation.

The reaction of Phil's co-workers reflects the ambiguity that festered within the Pontiac Street lab. This was truly a corporate laboratory. The environment was very unlike the atmosphere Phil had created around him in San Francisco and Philadelphia. In those days, Phil empowered all his men with the spirit that they were working *with* him, not *for* him, and they returned that spirit in the manner that they responded to his initiatives. Tobe Rutherford, Arch Brolly, Cliff Gardner, Russ Varian...they all understood where Phil was going, and often anticipated his next moves. Those men were always *right there with him.*

That kind of camaraderie was clearly absent at Pontiac Street. Maybe the work was easier in the earlier days. Maybe this nuclear stuff was just too far out on the frontier for anybody lacking Phil Farnsworth's unique vision to grasp. In any event, Phil never got the sense that these men were up to speed on the critical aspects of his approach to fusion. Yes, they seemed to work well *with* him, but the fact was they worked *for* ITT. They may have been co-workers on Phil's quest for fusion, but they were beholden to their employer, and Phil was never certain of ITT's motives or its commitment to his fusion work.

Whatever his misgivings about his working environment, Farnsworth continued to improve the Fusor through 1962 and 1963, seeing steady increases in the neutron counts with each modification.

Another remarkable event took place in the latter months of 1963, which has been reconstructed from interviews with members of the Pontiac Street team.[122] A verbal account from engineer Fred Haak described an occasion when he, George Bain, and another engineer named Jack Fisher were preparing the Fusor for a metered run that would be conducted the next day. There was no instrumentation on the Fusor during the setup. As was often the practice, the workers were putting the Fusor through its paces to make sure its systems were all functioning when, according to Fred Haak, the Fusor in the pit "just lit up and went crazy." George Bain killed the power immediately but the Fusor did not shut down—it actually continued operating, as an increasingly bright light emanated from the pit. After this spontaneous operation had continued for at least 30 seconds—perhaps a minute—a "pop and a hiss" indicated that the stainless steel reactor vessel had been breached, releasing its vacuum, at which point the reaction finally ceased and the Fusor cooled down.

Jack Fisher, one of the engineers on the fusion project, later related the story to Steve Blaising, another engineer who was not present at the time. Blaising was in charge of monitoring all the radiation-detecting equipment in the lab. That included "dosimeter" badges worn by all the personnel and also placed in strategic locations around the lab and in the actual Fusor pit. As soon as Blaising heard Jack Fisher's story, he grabbed all the radiation badges and sent them off for a reading. Typically, the badges would show almost zero exposure with each monthly reading, but this report came back indicating that the radiation badges were completely saturated, with reading levels completely off the scale. In other words, when the Fusor went off on its own on this occasion, the radiation readings suggest it produced substantially more fusion than on any previously measured occasion. Unfortunately, the lab's neutron counters were not in place, so there was no way of knowing exactly what level of neutron flux was attained, or whether the Fusor might have approximated breakeven.

What is most revealing about this incident is that, even though the reactor vessel was damaged and took some weeks to repair, the incident was not recorded, nor reported to any of the higher-ups at ITT. The workers were apparently afraid that the ITT management might arrive at the conclusion that the work was too dangerous and shut it down—a move that might have cost them their jobs.[123] Whether or not that fear was justified, the evidence certainly suggests that the Fusor was capable of much more than any of the written records would indicate.

What we do have are lab notes and records telling us that on February 7, 1964, the Fusor called "Mark II Mod 2" surpassed Farnsworth's immediate goal, recording a neutron count of 1.35 x 10^9. These notes also state that "the upper limit [was] fixed not by the capability of the device, but only by the radiation tolerance of personnel in the vicinity,"[124] again reflecting the inadequacy of the facilities that ITT was willing to provide. Nevertheless, at this level the operation was "completely stable and could be controlled and repeated."[125] These notes confirm that the Fusor was a very reliable device, that the experiments were very successful—and that the facilities were woefully inadequate for testing at the levels Farnsworth believed further success would require.

The curious confluence of these events lends some credence to the notion that the Fusor was just beginning to demonstrate its inherent potential—albeit in an occasionally unpredictable manner—during the winter of 1963–64. So it seems ironic that ITT would begin at this time to grow impatient with its nuclear energy project.

This impatience is most keenly expressed in ITT's insistence that Farnsworth add more "certified" personnel to the project, meaning an individual with more recognized credentials in the field of nuclear physics. We know that throughout this period, ITT kept looking to the Atomic Energy Commission for some kind of validation. The ITT management was well aware that the AEC was providing substantial funding for other lines of fusion research and felt as if ITT was having to use its private funds to compete with publicly funded experiments elsewhere around the country. Feeling that ITT was also entitled to some share of public money, management resolved to populate the operation with the sort of institutionally certified "experts" that the AEC seemed to favor in its funding.

Furthermore, the corporate executives who were being asked to finance this work clearly did not understand what they were getting into. Not only was nuclear energy well outside the purview of what was ostensibly a telecommunications and electronics company, but the research they were funding was well beyond their ability to keep up with Farnsworth's actual thinking. If they "got" it, it really didn't fit in, and even if it did fit in, they really didn't "get" it. Little wonder that the tenor of life at Pontiac Street began to change come the spring of 1964.

Farnsworth's advocate among the ITT management, Fritz Furth, tried to forestall the inevitable, but not even Furth could withstand the board's insistence that the project be given more scientific cachet. He implored Farnsworth to seriously consider adding some "lettered" personnel to his staff. Fortunately, a suitable candidate surfaced in June 1964 when Robert L. Hirsch arrived in Fort Wayne with a newly minted doctorate in nuclear physics from the University of Illinois. Hirsch brought with him a keen interest in electrostatic fusion, which he had investigated during his graduate studies, and a personal resolution to make fusion his "life's work."[126]

With the ink barely dry on his doctoral thesis, Bob Hirsch was determined to make a mark for himself with his first job. He wasted no time asserting his own interpretations of the Farnsworth approach to fusion and putting his own imprimatur on the Fusor design and experiments.

With Hirsch's arrival, "the Admiral," as Fritz Furth was fondly addressed, found himself juggling Hirsch's and Farnsworth's differences of method and personality. Farnsworth began to harbor suspicions regarding the motives of the brash young Ph.D. and felt that Hirsch was moving too fast in directions he was not really sure of.[127] Though their relationship appeared congenial on the surface, the underlying tension mounted when Hirsch had constructed, at some considerable expense, a concrete "cave" in which to conduct his own Fusor experiments. Some say that this development created an atmosphere of constructive competition within the workplace. Others say it produced an environment rife with friction and mistrust.

Farnsworth continued to conduct tests with the Fusor and by 1965 believed he was tantalizingly close to his elusive goal.[128] One nagging obstacle had the entire team stumped. Once the Poissor formed and began producing meaningful amounts of fusion, a layer of positively charged particles—a "virtual anode"—would form around the reaction, making it difficult for new fuel to find its way into the center of the reactor chamber. The various team members such as Gene Meeks, George Bain, and Bob Hirsch each had his own ideas about what was causing the problem and how they might get around it. Phil often looked at what his colleagues were trying and just shook his head.[129] Whatever it was, they just didn't get it, either.

Phil believed all along that the so-called problem of the virtual anode—this "force field" that kept new fuel from reaching the artificial star—was actually one of the predictable properties of the Poissor, an elegant controlling feature that naturally kept the reaction from running out of control or burning itself out. He had some ideas of his own about how the effect could be manipulated so that the formation of the virtual anode would not snuff out the reaction, but would in fact work to the advantage of the entire process. This was very much on his mind when he came home early one day. Pem noticed a curious look of suppressed excitement on his face. Phil was unusually quiet during dinner; Pem finally asked him what was on his mind.

"We made a run today. I didn't dare take it up as far as I wanted to, but I want you to see this."

Pem drove Phil back to the lab, which she described as dark and empty, in a moment that plays like a scene from a Steven Spielberg movie as they negotiated their way past the night watchman.

Once inside the dimly lit lab, Pem noticed that the Fusor itself was still resting at the bottom of its hydraulic pit where it had been tested earlier in the day. Phil gestured her toward a seat where she could observe the various meters that monitored the functions within the Fusor. Pem focused her attention on the meters, and Phil took his position at the power supply and started turning the rheostats that applied voltage and current to the Fusor. Very gradually over the course of several minutes, he increased the power.

Pem recalled, "Not knowing what to expect, and with chills running up and down my spine, I kept my eyes glued to the needle Phil told me to watch, reporting its position to him as it climbed the scale. At first, the needle settled in a position about three-fourths of the way up the scale."

Then Phil tweaked the power supply slightly, and the needle shot all the way up the scale, pegging itself at the extreme value. "It's gone off the scale," Pem called nervously to Phil.

At that, Phil quickly pulled back on the power. And then, "the most amazing thing happened: the needle just stayed right where it was. All power had been cut off to the Fusor, but that needle just stayed stuck…for a period of at least a half a minute after Phil had shut it off."

When the needle finally started fading back toward the zero mark, Pem turned to Phil, who was now staring at the meter as intently as she was. She waited for Phil to tell her what it meant.

Phil finally broke the silence, saying quietly, unemotionally, "That's it. I've seen all I need to see. Let's go home."[130]

After that unexplained event, Farnsworth began withdrawing from the day-to-day work around the fusion lab, practically becoming a spectator in his own project. Steve Blaising later recalled one occasion when Phil was at the lab, when some dignitaries from the Franklin Institute were on hand for a demonstration of the Fusor.

Farnsworth, the Admiral, Bob Hirsch, and others were gathered in Hirsch's cave lab that morning in 1965. The visitors asked questions about the principles on which the Fusor worked. "Phil started to explain inertial containment," Blaising recalled, "and used the expression 'electron return time.' One of the visitors cut him off in a very caustic voice and said, almost condescendingly, 'where do you get *that?*' And the Admiral said, 'Well, Phil has some expressions that are only meaningful to himself.'"

Before Phil could explain any further, the Admiral turned to Hirsch, saying, "Perhaps Bob better explain this to you." From his position near a control console, Steve Blaising watched as Hirsch took over the proceedings. Turning toward Phil, Blaising could see "the deep hurt in Phil. His head drooped, and his chin fell nearly to his chest in silence. If you ever kicked a dog, well, that's how Phil looked."

Farnsworth stood motionless and quiet for about one minute. Then, unnoticed by all but Steve Blaising, Phil stepped back to the rear of the cave room and slipped out the door.

About an hour later, the Admiral asked Blaising if he'd seen Phil. Calling the guardhouse at the gate, Steve learned that "Mrs.

Farnsworth had come by and picked him up a little while ago." Farnsworth had left the building.[131]

Phil Farnsworth returned to the Pontiac Street lab only one more time, a few days later. He assembled with Fritz Furth, a mathematician named Hans Salinger, and his patent attorney George Gust, and gave them what they needed to file the Fusor patent.

TWENTY

Tranquility Base

Farnsworth at home with Pem in Salt Lake City, ca. 1970

"God takes away the minds of these men, and
uses them as his ministers."
—Socrates

Like a half a billion of their fellow earthlings, Mr. and Mrs.
Philo T. Farnsworth sat in their living room in Salt Lake City,
transfixed in front of their television set as the Apollo 11 Lunar
Module "Eagle" began its final descent toward the surface of the
Moon on July 20, 1969.

Gene Kranz, the flight director, conveyed the decision of mission control to Neil Armstrong and Buzz Aldrin:

"Eagle, Houston. You are go to continue power descent."

Although his own involvement with television ended once
ITT sold off the Capehart-Farnsworth factories, Farnsworth always
kept a watchful eye on his brainchild as it swept across the nation and around the world. He never cared for most of the pro-

gramming he saw broadcast on his invention. His family retells the story of one time when he rose from his chair in disgust and emphatically switched off the set, beginning a period of years before a TV ever came on again in his household. He often felt that the medium's more constructive applications had been neglected, and wondered aloud at times if all the energy he'd spent on television was worth it.

"Eagle, Houston. You're go for landing."
"Roger, Houston. Go for landing."

It was a matter of little consequence to Philo T. Farnsworth that his own star faded just as the medium he pioneered was ascending to prominence in the world's cultural firmament. He was not one to dwell on past accomplishments. When he did reflect on the past, he was invariably determined to share the credit. On those occasions when he did speak in public or write about his contributions to television, he often started out by saying, "Years ago, my wife and I started television."

Beyond such infrequent reminiscences, Farnsworth's mind was always focused on some new vision of the future. The primary reason he was interested in fusion was for space exploration. He always believed he was going to travel in space, and predicted as much to his wife as they stood on the deck of a ferry crossing the San Francisco Bay one starry night in 1926. He never did get to make a space flight himself, but it pleased him to see others beginning the journey during his lifetime.

"Houston, we're go. Hang tight. 2,000 feet, 47 degrees."

There are only a few noble spirits like Philo T. Farnsworth who appear on Earth during each generation, men and women who can alter the course of history without commanding great armies. Albert Einstein was one. Nikola Tesla, who invented alternating current and battled with Thomas Edison over it, was another. Edwin Armstrong, who invented the circuits that made broadcasting possible, also fought with David Sarnoff over his patents for FM radio until his death.

Such gifted individuals seem to arrive with a special insight embedded in their soul, some unique, inexplicable grasp of nature that not only creates fabulous gadgets but also stretches the frontier of human understanding. It seems that no matter how noble these spirits might be, their presence is often impeded by

Farnsworth at his property in Maine

less-enlightened forces here on Earth, who unwittingly oppose them, sometimes in the very name of progress. At times we are fortunate to receive even a fraction of their gifts.

"750, coming down at 23...700 feet, 21 down...."

Phil tightened his grip on Pem's hand as the Eagle descended ever closer to the lunar surface.

As Albert Einstein was to Isaac Newton, might Phil Farnsworth have been to Einstein? Was Farnsworth on the threshold of a new level of human understanding somewhere beyond Einstein's universe? As Einstein had stood on the shoulders of Newton and reached for the next rung of knowledge, was Farnsworth standing on the shoulders of Einstein, hoping that his earthbound star would catapult mankind into the heavens? Had he discovered in the power *of* the stars an idea that would enable travel *to* the stars?

We can't answer that question because Farnsworth never followed through on his promise to Einstein to publish the math. At the same time he was trying to formalize the math, he was struggling with the Fusor. When his colleagues questioned that math, he focused on the Fusor as a means of proving his theories. He knew he was traveling in uncharted territory, and he struggled daily with the difficulty of conveying his ideas across the confounding distance from his grasp to the reach of others. He expected the Fusor would silence his detractors, merely by working as he predicted. When he realized he would be denied the chance to finish the work, he had only enough time and energy left to satisfy himself.

Farnsworth often said his life was a "guided tour." The native instincts embedded in his soul's code led him to the ideas that gave the world electronic television. Working on television, he encountered the phenomenon of secondary electron emissions. Working with secondary emissions led to the invention of the Multipactor tube. The Multipactor, in turn, revealed the secret that may have solved the riddle of fusion, a clean, safe, and limitless source of energy for our planet. We are left to wonder why this noble spirit could not ultimately deliver his most inspired creation.

"300 feet, down 3-1/2. Got the shadow out there."

We are left to ponder the titanic struggle between the forces that inspire mankind to reach its destiny among the stars and those that keep it anchored to the Earth.

Farnsworth always believed we would travel in space. Whether or not he truly believed it would happen to him—that he would personally have an opportunity to leave Earth's gravity and explore the heavens—he knew that to be our destiny. And he knew, perhaps from birth, that he had a role to play in achieving that destiny. His unfinished work remains just that: unfinished.

And so the riddle of fusion becomes the riddle of the Farnsworth Fusor: Did it work, or was Farnsworth's approach a dead end? Nothing in the record gives us a conclusive answer either way, so the mystery persists to this day. Asked to speculate why the Fusor project ended so inconclusively, Bob Hirsch said simply, "Not enough money."[132]

"75 feet... looking good... picking up some dust... 30 feet, 2-1/2 down... Contact light. Okay, engine stop... Houston, Tranquility base here. The Eagle has landed."

"Roger, Tranquility, we copy you on the ground. You've got a bunch of guys about to turn blue. We're breathing again. Thanks a lot."

Like the rest of the world at that moment in 1969, Phil and Pem Farnsworth started breathing again, too. They smiled knowingly at each other when the near-speechless Walter Cronkite managed to regain his composure long enough to comment that, amazing as the lunar landing itself was, even more amazing was the fact that the entire world was sharing the event through television.

Of course, the world didn't watch the actual landing—they listened to the live audio transmissions between CapCom and the Lunar Module, and on some networks they watched animated simulations of the landing. Nobody got to witness the landing itself because no live television cameras were deployed until the astronauts were ready to step out of the lander.

Several hours later, Armstrong and Aldrin suited up to leave the Lunar Module and walk on the surface of the Moon. As Armstrong waited on the edge of the ladder, Aldrin flipped a circuit breaker and radioed to Houston.

"Roger, TV circuit breaker's in. Receive loud and clear."

At home in Salt Lake City, the inventor of electronic television caught his breath again. He had learned through his sources in the industry that the camera that NASA had sent up with Apollo 11 used a tube that was based on Farnsworth's original Image Dissector, the tube that "had everything that it needed and nothing that it didn't."

"Man, we're getting a picture on the TV...Okay, Neil."

A few moments later, Neil Armstrong took his historic "giant leap for mankind" as he stepped off the footpad of the Lunar Module onto the surface of the Moon. A quarter of a million miles away, television turned one man's lunar stroll into an expression of global awakening, a moment in which the entire planet became involved in the unfolding of its own evolution.

For Philo T. Farnsworth, this extraordinary occasion provided a long-delayed moment of personal triumph, which erased any doubts about the value of his contribution. Just seeing with his own eyes that his invention made it possible for the entire world to witness those historic steps was enough to make him turn to his wife and say:

"This has made it all worthwhile."

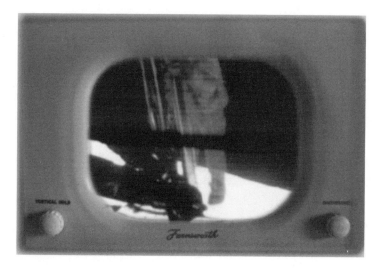

The Sword in the Stone

Farnsworth portrayed in an ITT promotional sketch

"The most beautiful thing we can experience
is the mysterious."
—Albert Einstein

Despite Farnsworth's abrupt departure from the patent meeting in 1965, patents were issued for the Fusor in 1968. Like the fusion work itself, the patents have been described as "incomplete," largely because of Farnsworth's abandoning the patent-writing process. When he realized that even his most trusted and knowledgeable colleagues still didn't "get it," Farnsworth packed up his papers and went home. He got good and drunk, and determined to stay drunk for as long as it took him to leave behind his mortal coil. Over the next few weeks he descended into a familiar spiral of despair that almost killed him directly.

Farnsworth's employment with ITT ended when the company retired him for medical reasons in 1966. Nevertheless, with help and inspiration from his family, he recovered from this descent, as he had from others. He pulled himself back together, and in 1968 some of his most loyal co-workers followed him to Salt Lake City, where he made one more effort to regroup. Phil and his small team hoped to make another attempt to "light the fusion torch." Oddly, despite ITT's professed lack of faith or interest in his work, Farnsworth found his former employer unwilling to release the patents or even sell them to him. He formed a new company, Philo T. Farnsworth and Associates, to pursue a number of other projects. The venture seemed to be going well until a shady promoter pulled the financial rug out from under him, ultimately costing him what little fortune he had left from a lifetime of seminal research.

From there his health once more began to fail, and Farnsworth died in Salt Lake in March 1971 at the age of sixty-four. The official cause of death was listed as congestive pneumonia, but the family says he died of a broken heart.

So ended the life of Philo T. Farnsworth, devoted husband, father, scientist, and inventor. But his spirit lives on and the fire of his genius is re-ignited billions of times every day—every time a television set is turned on.

For the past thirty years, the Farnsworth family has been reluctant to talk about Farnsworth's fusion work. Pem Farnsworth has always expressed a certain reverence for the metaphysical issues that her husband struggled with during the years he was trying to perfect the Fusor.

When they returned to Salt Lake in the late '60s, Phil and Pem restated their marriage vows in a sacred ceremony inside the Mormon Temple—a ritual intended to ensure that their souls would be reunited in the hereafter.[133] Pem is a deeply religious woman, and an underlying expression of her faith is evident whenever she talks about Phil's fusion work. "The time just wasn't right," she has said, although she also feels strongly that, in their late-night visit to the lab, Phil was afforded a moment on the mountaintop—and a validating glimpse of the promised land that he would not be permitted to enter.

Phil was, on a gut level, always apprehensive about the impact that fusion would ultimately have on civilization. He never stopped worrying that man might not be

Pem and Phil after their
Mormon Temple wedding in 1970

spiritually mature enough to have such immense power at his disposal. For these reasons, family members most familiar with the story have said on more than one occasion, "He took the secret to his grave."

In 1967, Robert Hirsch tried to continue the fusion work at ITT without Farnsworth. Hirsch and Fritz Furth submitted an exhaustive proposal to the Atomic Energy Commission, requesting less than half a million dollars to continue their experiments with Farnsworth's Inertial Electrostatic Confinement. The proposal was accompanied by numerous endorsements from experts in the field who, to a man, added their opinion that the approach was viable and worthy of further research.

Late in 1967, Bob Hirsch and Gene Meeks flew to Washington with a specially built Fusor, which they mounted on top of a hotel dessert cart—the first true "desktop" fusion reactor! They rolled their portable Fusor into a meeting of the Atomic Energy Commission, literally plugged it into a wall socket, and produced fusion for the assembled officials, all of whom represented some previously funded fusion project.

Hirsch recalled the event with some irritation. "We built something incredible. We rolled it into the Atomic Energy

Commission and plugged it in to the wall, and it was just amazing to these guys. They were trying to figure out what was wrong, like they were looking for the man behind the curtain. They looked in and there's this bright spot in the middle and it just stared right at them, and the neutron counter went click-click-click and that was kind of mind-boggling to them."

What followed gave Hirsch his first good taste of how decisions are made within the scientific bureaucracy. "I thought I did as good a job as I was capable of doing," Hirsch remembered. "I was answering questions honestly. When I didn't know I told them I didn't know. I told them what I thought and what was speculation, and how the experiments were run and everything. I thought it was going very, very well. The turning point was when a fellow named Tom Stix, who was a professor from Princeton, leaned back in his chair, and said, 'If we fund this, whose budget is it going to come out of?' I will never forget those words as long as I live because that's when I knew the game was over."

In a January 1968 letter, the AEC formally notified Hirsch and Furth that there would be no federal funding for the proposed ITT fusion research program, citing insufficient funds. The AEC chose instead to place its bets on the horse called Tokamak, and in the past thirty years has spent billions of dollars on massive machines that to this day have still not produced a fraction of the fusion that Farnsworth produced in the 1960s with his simple Fusor.

That Farnsworth's work on fusion ended so inconclusively remains baffling to Hirsch, who left ITT in 1968 and joined the Atomic Energy Commission—the same agency that had denied funding for his continued research on electrostatic fusion. When the AEC became part of the U.S. Department of Energy, Hirsch rose to the level of director of the national fusion program, where he served from 1974 to 1977. From his vantage point at the top of the entire nationwide fusion effort, Hirsch gained a unique, firsthand perspective of the rigid orthodoxy that shapes the country's energy research policies.

Throughout his tenure with the DOE, Hirsch was frustrated in his attempts to secure adequate funds for electrostatic fusion research. Asked to describe his experience, Hirsch offered this assessment of the scientific establishment:

"The scandal in the whole thing," Hirsch said, "and the thing I don't understand to this date—and maybe never will—[is] how those people would not open up to the

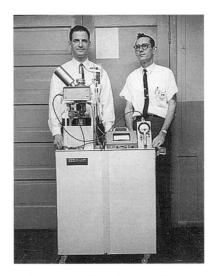

Robert L. Hirsch and Steven Blaising with the "dessert cart" fusor, ca. 1967

possibility of a Farnsworth-like idea. Good physics and good research should be done with a relatively open mind. When you're dealing with a complicated problem you should not over-constrain yourself and you should look in a lot of different corners to understand what the possibilities are. The people in the program were almost paranoid when it came to this particular subject."

On the specific subject of Philo Farnsworth, Hirsch added, "I think they were probably also uncomfortable with Farnsworth, in the sense that here was an inventor, a farm boy with just dribs and drabs of education, who in fact conceived and developed one of the most significant technological advances of the 20th century, and here he was coming along in fusion, and I don't know whether it was ego or what but there is something strange there...."

The patents issued in 1968 expired in the mid-1980s, and whatever original art they disclosed is now in the public domain. If anybody can figure out, based on those patents, what was really going on inside the Fusor, we may yet find a door to the abundant future that Philo Farnsworth could only describe for us

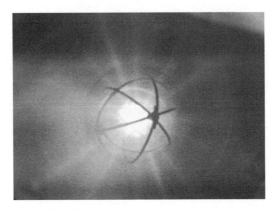

Dozens of amateur scientists around the world have
achieved fusion, like in this Fusor built by Richard Hull

from his vantage point atop the summit of his unique knowledge and insight.

Like the mythological sword Excalibur, the secret of the
Fusor lies frozen in the rock of institutional orthodoxy. A few
nuclear physicists around the world are beginning to experiment
with the Fusor, but their funding is still inadequate. There is also
a small, dedicated cadre of scientific amateurs who are building
their own small Fusors. Owing to the inherently simple and
straightforward design of the Fusor, these students and weekend
scientists are actually producing low-level fusion reactions in
their basements and garages. They add daily to a growing body
of knowledge of the subject.[134] Their work lends credence to the
prospect that what has not been delivered at the institutional
level might one day arrive from some unlikely place down at
the grassroots—from the very same fertile intellectual soil that
nurtured the seed of electronic television.

Like the knights and squires of the ancient legend, these
would-be kings can stand on the rock and try with all their
might to pull the sword from the stone. But when the time is
right, the mystery will finally reveal itself, the stone will unlock
its secrets, the quiet passion of Philo T. Farnsworth will be re-
awakened, and the dawn of a new civilization will truly be
upon us.

Who Invented Television?

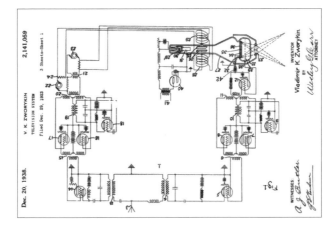

Zworykin's 1938 Iconoscope patent, with the 1923 application date

"I know that God exists. I know that I have never invented anything. I have been a medium by which
these things were given to the culture as fast as the
culture could earn them. I give all the credit to God."
— Philo T. Farnsworth

As compelling as the story of Philo T. Farnsworth may be, the
historical record with regard to "who invented television"
remains fuzzy at best, deliberately distorted at worst. The debate
often comes down to a simple question: Does any single individual deserve to be remembered as the sole inventor of television?
Can we create for television the kind of mythology of individual,
creative genius that history has bestowed on Morse, Edison, Bell,
or the Wright Brothers?

The question may be simple, but clearly the answer is not.
Before Uncle Milty, before Walter Cronkite, before Lucy and Desi
and Ethel and Fred, literally hundreds of scientists and engineers

contributed to the development of the appliance that now dominates "our living room dreams." How can we single out any single individual and say, "it all started here"?

The historical record is sadly devoid of references to Farnsworth. Though the oversight has begun to improve in recent years, it is still entirely possible to open an encyclopedia and read that electronic television began when "Vladimir Zworykin invented the Iconoscope for RCA in 1923..."—a sentence that manages to express no less than three historical inaccuracies. The most conspicuous error—the "1923" date—fixes Zworykin's name chronologically before Farnsworth's 1927 patent filing, and often renders Farnsworth to the status of "another contributor" in the field.

Some historians have gone so far as to suggest that Farnsworth and Zworykin should be regarded as "co-inventors." But that conclusion ignores Zworykin's 1930 visit to Farnsworth's lab, where many witnesses heard Zworykin say "I wish that I might have invented it." Moreover, it ignores the conclusion of the patent office, in its 1935 decision in Interference #64,027, which states quite clearly "priority of invention awarded to Farnsworth."

These misinterpretations of the historical record are precisely what more than sixty years of corporate public relations wants us to believe—that television was "too complex to be invented by a single individual." But close examination of the stories beneath the written record reveals a far more compelling story: In fact, there *was* one inventor of electronic television. Video as we now know it first took root in the mind of Philo T. Farnsworth when he was fourteen years old, and he was the first to successfully demonstrate the principle, in his lab in San Francisco on September 7, 1927. If you need to fix a date on which television was invented, that's the date.

Before that date, television was the province of Newtonian electro-mechanical engineers who employed spinning disks and mirrors in their crude attempts to scan, transmit, and reassemble a moving image. The inventions of Jenkins, Ives, Alexanderson, Baird, and others are all similar in their reliance on the spiral-perforated, spinning disk first proposed in the 1880s by the German Paul Nipkow. These contraptions were engineering marvels in their own quaint way, but they were not the sort of breakthrough that Farnsworth introduced, nor is anything left of their technology in the system of television that is in use around the world today.

On September 7, 1927, Philo T. Farnsworth demonstrated for the first time that it was possible to transmit an "electrical image" without the use of *any* mechanical contrivances whatsoever. In one of the first triumphs of Relativistic science, Farnsworth replaced the spinning disks and mirrors with the electron itself, an object so small and light that it could be deflected back and forth within a vacuum tube tens of thousands of times per second. Farnsworth was the first to form and manipulate an electron beam, and that accomplishment represents a quantum leap in human knowledge that is still in use today.

After September 7, 1927, every new contribution to the art—including Zworykin's—was an improvement on Farnsworth's simple, elegant, and profound invention.

What is so often overlooked cannot be overstated: In 1923, Vladimir Zworykin—recently emigrated from Russia, and employed at the time by the Westinghouse Corporation in Pittsburgh, Pennsylvania—*applied* for a patent for an approach to television that he first encountered in the classroom of Boris Rosing, his former teacher in Russia. In 1927, Farnsworth also applied for a patent. Later that year, Farnsworth produced the first successful transmission of a television image by wholly electronic means—an event that is thoroughly documented in Farnsworth's journals—while Zworykin's application was still pending. Farnsworth's patent #1,773,980—with the critical Claim 15 regarding the "electrical image"—was issued in August 1930—and Zworykin's application was *still* pending.

The 1923 Zworykin application would be forgotten—except that a patent for the Iconoscope was finally issued in 1938 bearing a 1923 application date. This patent (#2,141,059) was issued an extraordinary *fifteen years* after the application date, and then only after extensive revisions had been made to the original application.

Furthermore, the eventual patent granted pursuant to the 1923 application was issued over the objection of the patent office, and even then not until the case was adjudicated by a court of appeals. That the Iconoscope patent was issued at all hinged on a technicality, and it served no practical purpose other than substantiating the dates that RCA would eventually use in its public relations campaign.

RCA's obtaining the patent in 1938 has served as the cornerstone of its efforts to influence the historical record, since the patent effectively fixes 1923 as the date that Zworykin first disclosed electronic television. Decades later, historians and scholars are still including this dubious 1923 date in their chronologies.

What's wrong with the Zworykin patent? What's wrong with it is that the original application—the system that Zworykin disclosed in 1923—simply could not work. The idea was on the right track, but the application fell far short of disclosing a device that would pave the way to electronic video and ultimately put a television in every living room or a computer monitor on every desktop.

There is scant evidence that Zworykin ever built and tested a system like the one disclosed in his 1923 application. One story does exist about Zworykin's attempt to demonstrate his concept for executives of Westinghouse, where he was employed at the time, in hopes of obtaining more funding for his research. The demonstration was so dismal that instead of providing him with further funding, Zworykin's superiors ordered him to find something "more useful" to work on.[135]

The usual retelling of this story is cast in such a way that we are supposed to believe that the Westinghouse executives who witnessed and dismissed this demonstration were too shortsighted to appreciate its promise. It seems more plausible to conclude that what they saw showed little promise because it simply didn't work. Some historians suggest that witnesses observed some sort of blurry smudge. Zworykin would claim years later that the image of a cross was transmitted. But during the critical 1934 interference proceedings there was no evidence submitted to support even these modest contentions.

It's hard to imagine anyone in 1923 or 1924 seeing even an incoherent transmitted image on the bottom of a bottle and telling its creator to find something "more useful" to work on. But that's what we're supposed to believe.

The most recent accounts of Zworykin's debatable patent history are often traced to *The History of Television: 1884–1941* by Albert Abramson. A careful examination of Abramson's book only serves to further illustrate the flimsiness of this account.

The actual evidence that such a demonstration ever took place is sketchy at best, considering its potential historical significance. There are no lab notes, no direct eyewitness testimony. There are only Zworykin's own accounts, and a *single* document on page 80 of Abramson's book that he claims to have found buried in some archives fifty years after the purported event. This document describes a device "using a modified Braun type cathode ray tube for transmitter and receiver…the receiving tube …gave quite satisfactory results…[but] the transmitting part of the scheme caused more difficulties…."[136]

That's it; that's all it says about the transmitter, that it "caused more difficulties." It's hard to imagine how the receiver could be "quite satisfactory" if the transmitter was not equally satisfactory, but this is the document that compels Abramson to conclude—in his footnotes—that "Zworykin did build and operate the first camera tubes in the world sometime between the middle of 1924 and late 1925."[137] This is the feeble foundation on which historians build RCA's claim that Zworykin should be regarded as the "inventor of television."

Zworykin may indeed have built some tubes. And he may have applied current to them. But it should take more than a statement that "the transmitter caused more difficulties" to convince students of this history that he successfully "operated" such a device prior to September 7, 1927, or that Zworykin even deserves to be considered a "co-inventor" as a result of this experiment.

Historians should focus more carefully on the decision of the U.S. Patent Office in its historic 1935 ruling in Patent Interference Number 64,027. This is the litigation in which Zworykin challenged Claim 15 in Farnsworth's patent #1,773,980, which describes the "electrical image." An electrical image is the electrical counterpart to an optical image. When an optical image is focused on a photoelectric surface, the light-sensitive chemicals emit an array of electrons—the "electrical image"—which can then be scanned to form a fluctuating current. That is the very essence of how an electronic television signal is created, and so it is understandable that Zworykin and RCA would attempt to appropriate the language in this claim. There is simply no getting around it—you can't create an electronic television signal without first creating an "electrical image."

The whole of RCA's research effort—at an expense that David Sarnoff joked with Zworykin years later cost RCA more than $50 million—was intended to circumvent Farnsworth's patents, in particular Claim 15. When the electrical image in Claim 15 proved essential, Sarnoff, Zworykin, and RCA's attorneys went to great lengths in the 1934 interference to prove that the 1923 application would have created such an electrical image, and that Zworykin was therefore entitled to "make the count" embodied in Claim 15.

But when it was time for RCA to produce material evidence that Zworykin had constructed and operated his system in 1923, there was no evidence submitted. No tubes were displayed, no laboratory journals entered into the record. There were only confusing and contradictory verbal accounts from two Zworykin colleagues.

After considering all the testimony, the patents examiners ruled in Interference #64,027 that "Zworykin has no right to make the count because it is not apparent that the device would operate to produce a scanned electrical image unless it has discrete globules capable of producing discrete space charges and the Zworykin application as filed does not disclose such a device."[138]

The patent examiners were unequivocal in their decision to award the indispensable Claim 15 to Farnsworth. The case was appealed and RCA lost all the appeals. This pattern went on, over this and other patents, until RCA capitulated in 1939 and accepted a license from Farnsworth for the use of his patents—the first such license in the history of a company that was determined to "collect patent royalties, not pay them."

Yet, here we are nearly seventy years later, still debating the merits of a patent that was awarded by a court of appeals in 1938 that validated a patent applied for in 1923 that was ruled inoperative in 1934.

The contradictions are clear: What we have is an application for a patent in 1923, an unsuccessful demonstration in "1924 or 25" with no conclusive documentation, and a patent interference ruling in 1934 that says the device was inoperative. Nevertheless, a patent was obtained in 1938 which compels otherwise scholarly reporters to conclude that Zworykin and Farnsworth must be considered co-inventors.

A more discerning examination of the record reveals that Zworykin believed in electronic television but was still struggling for a viable solution until he visited Farnsworth's lab in 1930. As soon as he saw what Farnsworth had achieved, he got busy, duplicating Farnsworth's equipment at the Westinghouse lab in Pittsburgh before moving on to RCA in Camden. He then built on Farnsworth's work, as well as the work of other contributors, to produce the Iconoscope.

Zworykin's corporate benefactor, David Sarnoff, believed the Iconoscope gave him the leverage he needed to bring all the legal might of RCA to bear on claiming Farnsworth's achievement as RCA's own. Sarnoff ultimately failed in that effort, and RCA was left with no choice but to accept a patent license from Farnsworth. Still we read time and again that Zworykin made modern television possible when he "invented the Iconoscope for RCA in 1923." The facts are that Zworykin was not working for RCA in 1923, the Iconoscope did not exist at that time, and it is questionable whether Zworykin truly invented it at all.

Zworykin got some momentum going with the Iconoscope, but it was not until the Image Orthicon tube was introduced that the industry had the tool it really needed to bring the world into our living rooms. But the Image Orthicon—originally thought to be an RCA development—was in fact descended from Farnsworth's patent #2,087,683, which was the first to disclose a "low velocity" method of electron scanning. This lends further credence to the notion that everything that came after September 7, 1927 was an improvement on the concept proven that day—including Farnsworth's own subsequent inventions.

That said, there is no question that much credit for refining all aspects of television technology goes to RCA engineers. There were hundreds, maybe thousands, of individuals who contributed to the development of electronic video before television broadcasting reached the general public in the 1950s, and thousands more who have contributed to its advancement in the decades since. But refinement is not invention, though that is precisely what the proponents of the "co-inventor" theory of the origins of television would like us to accept.

Why is any of this important? Who really cares who invented television? What difference does it make whether electronic television was first developed by a Russian émigré or a Mormon farm boy? And should it still matter seventy years after the fact?

It matters because the suppression of the true story deprives us of some important knowledge of the human character. It tempts us to believe that progress is the product of institutions, not individuals. It tempts us to place our faith in those institutions, rather than in ourselves.

Invention is one of the most unique and compelling aspects of the human experience. From the moment the first ape picked up a bone and swung it like a club, the history of civilization has followed the path of invention.

Szent-Gyorgi put it best when he said, "Discovery is seeing what everybody else has seen, and thinking what nobody else has thought." Therein lies the operative definition of genius. In Zworykin, we find a capable engineer, one who could see what others were doing and improve upon it. But in Farnsworth, we encounter the rarest breed of all, the true visionary who could see the obvious—and think up something entirely different. Obscuring his story and denying his contribution deprives us of our understanding of this critically important facet of the human character.

Television is our blessing and our curse. The ancient dream of a unified planet came true with the moonwalk in 1969, as hundreds of millions of people around the world tuned in to witness the event through the medium of Philo T. Farnsworth's potato-field inspiration. At the other extreme there are the routine daily programs that cater to "the lowest common denominators" of our society. But even these daily panderings to common culture are somehow elevated when reconsidered with the knowledge that the medium itself is a consequence of individual genius rather than corporate engineering.

The belief that television—the most pervasive mass communications system of the past millennium, and perhaps the next— was "too complex to be invented by a single individual" deprives us of the knowledge of the noble individual whose unique intellect made it all possible. There are only a few such souls in each century, men like Tesla, Armstrong, and Einstein whose lives are an enduring expression of Szent-Gyorgi's axiom.

Philo T. Farnsworth was as noble a spirit as has ever graced this planet. From his earliest declaration of his hope that he, too, had been "born an inventor" it is clear that this earthly soul was an instrument of providence. When he saw how the mad scientists of the 19th century tried to send pictures through the air with spinning disks and mirrors, he alone replaced all the moving parts with the invisible electron. Recalling that contribution makes even the most ordinary moments of television programming an expression of divine inspiration.

The Story of The Book

Paul Schatkzin with George Everson in 1975

"Genius is eternal patience."
— Michelangelo

This book has taken more than twenty-five years to finish.

I first heard of Philo T. Farnsworth in the summer of 1973. I had just graduated from a branch of Antioch College near Baltimore, where I majored in communications. I was off to California to seek my fortune in the television industry, when I happened across the most recent issue of a publication called *Radical Software*—an obscure underground periodical that promoted the use of portable video recorders for social action. Originally published from New York City, this new issue was published from San Francisco and called the "Videocity" edition, in homage to the medium's origins. The issue included an elegy to Farnsworth, "The Electromagnetic Spectrum Blues" by Max Crosley.

Lamenting the lack of TV coverage of Farnsworth's death on March 12, 1971, Max wrote:

Athena weeps,
The electro-magnetic spectrum has the blues,
And not one of you has been unaffected by this man.
THEY OWE IT ALL TO HIM . . . but they never said a word.
HE GAVE ALL
As they nothinged him right into nothing.
When Picasso was wildly experimenting with deco cement
THIS MAN WAS DRAWING WITH ELECTRONS,
You all know he went through here,
Whether you know his name or not.
HE RIGHT NOW IS COMMANDING YOUR LIVING ROOM
MIND.[139]

Until that chance encounter in the pages of a rather obscure publication, I had never given the origins of television much thought. But there was a quality to the illustrations that accompanied the poem, the photos of the young Farnsworth holding his tubes and standing by a crude wooden camera box in the 1920s, that filled in a piece of history that I did not even know was missing.

A few weeks later, rummaging through the stacks at the Santa Monica library, I stumbled across *The Story of Television: The Life of Philo T. Farnsworth* by George Everson. Everson, I learned from this book, was the man who discovered Farnsworth in 1926, helped arrange his financing, and remained one of his primary supporters until the company was restructured in the late 1940s. I read the book quickly, from cover to cover.

Still later that same summer, my former college roommate, Tom Klein (now a Hollywood TV producer) and I took a trip up the Pacific Coast Highway and stopped in Santa Cruz to visit with a fellow I remember only as "Johnny Videotape"—a pseudonym he'd adopted for his public-access cable video work. Johnny knew Phil Geitzen, who had edited the "Videocity" issue of *Radical Software*, and Geitzen knew Philo Farnsworth III, the TV inventor's oldest son. It was through this chain of connections, on a hill overlooking the Pacific in Santa Cruz in the summer of 1973, that I first heard the word "fusion." The apocryphal way Johnny Videotape

conveyed the story sent chills through me that has kept me connected to this story for more than two decades.

For some reason, I guess I've always had a thing for inventor
stories. When my mother was trying desperately while I was in the
third grade to get me to read, she took me to the library and I
picked out a Signature Series biography of Thomas Edison. The
next year, when we all got to pick a character to portray in the
fourth grade play, I chose to play Edison, and attempted to invent
the lightbulb in front of the entire Forrestdale School in Rumson,
New Jersey. Unfortunately, my lightbulb did not work quite as well
as Edison's. But a seed was planted when I read my first book
about Edison; a seed that took root that summer when I stumbled
upon these stories of Philo T. Farnsworth.

In the spring of 1975, I'd found work with Videography, Inc., a
small video production house in Hollywood. I was in charge of
promoting the company's computerized video editing services, and
suggested to the owner, Bob Kiger, that we issue a "video buck" to
our prospects, each coupon worth one free hour of editing. In
place of George Washington on our facsimile dollar bill, I suggested we place a portrait of Philo T. Farnsworth.

"Who is Philo T. Farnsworth?" Bob asked, half laughing at the
name.

I explained, "He's the father of television...he invented it..."
and proceeded to lay out the story I'd read in Everson's book of
the farm boy who had dreamed up electronic video.

"That's a great story," Bob said, "that would make a great
movie for television!"

With that seemingly reasonable suggestion, I was off on an odyssey that has woven in and out of my life for more than twenty-
five years, which continues even with the completion of this volume.

Bob and I tracked down George Everson—by then well into
his nineties but still quite coherent—in the rugged foothills of
Mendocino County, and we acquired an option on his book. From
there, we located the Farnsworth family in Salt Lake City and made
similar arrangements for the movie rights to the then-unfinished

book that Farnsworth's widow, Pem, was researching and writing at the time. We never did get a movie made for television, but the Farnsworth family has remained a near constant presence in my life ever since.

The material in this book is drawn largely from interviews I conducted with Pem Farnsworth and Philo T. Farnsworth III between July 1975 and September 1977. Those interviews, and supporting material I dredged up in the stacks of the UCLA Research Library, were first compiled into the treatment that Bob Kiger and I used in our efforts to interest the television networks in our project. Bob dropped out of the project in 1976, but my then-future-ex-wife, Georja Skinner, and I continued to carry the torch.

In 1977, in concert with the 50th anniversary of Farnsworth's first successful electronic video experiments, the material I had gathered was published in another obscure "alternate media" journal based in Washington, D.C. called *TeleVisions*. I worked closely with the editor of that publication, Nick DeMartino, to clean up the narrative I'd written for the TV treatment to make it suitable for publishing in four installments through the course of the anniversary year. The celebration culminated with a re-enactment of Farnsworth's successful September 7, 1927 experiment that was covered by two of the three major networks in their nightly newscasts.

The publication of that material in *TeleVisions* was the first effort of any consequence to compare the Farnsworth family's recollections with the existing historical record, much of which had been dictated over the course of the previous four decades by the public relations departments of those companies that survived the shakeout from television's early years. It has been suggested by some observers that the reclaiming of Philo T. Farnsworth's true legacy began with the 50th anniversary celebration and the publication of those four installments in *TeleVisions*.

It's nice to know that I had some role in setting the record straight. In the two-and-a-half decades since, a number of publications and media productions have basically confirmed and echoed the themes first expressed in those four *TeleVisions* installments. Most notable among them is Pem Farnsworth's own book, *Distant*

Vision: Romance and Discovery on an Invisible Frontier. When it was published in 1990, *Distant Vision* culminated fifteen years of Pem's own research and writing on the subject. I was privileged to work closely with Pem and her son, Kent Farnsworth, in the completion and publication of that book, and added substantially to my knowledge of the subject matter as a result of that experience. I recommend Pem's book to any reader who desires a more

Pem Farnsworth in 1977

intimate, firsthand interpretation of the material contained in the pages of *The Boy Who Invented Television.*

Another volume that is indispensable for any serious student of Farnsworth's life is *Philo T. Farnsworth: Father of Television,* written by Dr. Donald Godfrey, a professor at the Walter Cronkite School of Journalism and Telecommunications at the University of Arizona, and published in 2001. Godfrey's book is the most extensively researched and documented volume on the subject of Philo T. Farnsworth yet published, and sets the standard for journalistic excellence by which all future efforts must be judged. I was pleased that Dr. Godfrey saw fit to draw a little bit from the material I had published some twenty years earlier. His scholarly research largely confirms the stories I first wrote about in the 1970s.

I have taken the liberty of drawing on both of these works in the rewriting of my original treatment for this book. Wherever I have used direct quotations and excerpts from these books, as well as others, I have included reference notes. In doing so, I wish to recognize to the scholarship displayed by these authors, who have likewise acknowledged my own earlier contributions. I would like to think that my work stands on the shoulders of theirs, and by so doing extends the thematic reach of all the material now in circulation, which gets to the heart of the real issues that shaped the life of Philo T. Farnsworth, his successes and his shortcomings.

I never met Philo Farnsworth II, the inventor recalled in these pages. He died two years before I ever heard his name. I did, however, become very close to his oldest son, Philo T. Farnsworth III. Much of my understanding of the broader themes that challenged the elder Philo, I absorbed during the time that I was privileged to spend with "P3" in the late 1970s. Philo III was cut from the same cloth as his father, but lived his life in a manner much like

Philo T. Farnsworth III in 1972

a mirror image. Philo III was a reluctant inventor in his own right, reluctant because of what he had seen the process of invention do to his father. From his experience, I came to understand the eternal clash between invention and industrial capitalism, and the impact that had on the health, wealth, and well-being of both father and son.

In 1987 Philo III left this world through the same dark corridor that accounted for many of the detours in his father's life.[140] I sorely miss him. But I am fortunate that much of the time we shared together was recorded on audiotape. I returned to those tapes during the rewriting of this book and discovered a rich vein of material that had been necessarily overlooked in the original preparation of what at the time was a movie treatment. Listening to the interviews I conducted with Philo III and Pem during the seventies was like opening a time capsule. Where Pem provided most of the "play by play" based on her personal recollections of the events she experienced with her husband, Philo III provided all the "color commentary." His attitudes, philosophies, vocabulary, and vision infuse practically every paragraph of this text. Given the similarities of his character to that of his father, I would like to think that this book is how the story might have come across had his father had the opportunity—or the desire—to tell it himself.

Wherever possible, the stories in these pages are conveyed precisely as Pem and Philo III relayed them to me. I have cross-referenced their accounts with my own additional research and the work of Godfrey, Everson, and others, always striving to provide as much reliable historical accuracy as possible. It has always been important to me to create a narrative based on fact, with as little "documentary bio-drama" as possible. I don't want readers—or viewers, if ever there is a movie made—to sit and wonder if a particular scene really happened. I have tried not to invent scenes or storylines. All the scenes portrayed in this book really happened, to the extent that memory and ancient texts can reliably recall. I have taken the liberty of embellishing some scenes with suggestions of dialog and action, but the scenes themselves are derived from first or secondhand accounts of the actual events.

My own material languished in the years after its first publication in 1977. A second effort at organizing a feature film or movie for television after the publication of Pem's book suffered the same fate as the first effort in the 1970s. Then in 1995, as the Internet started bubbling into consciousness, I started an Internet business. I remember waking up one Saturday morning in 1995, the week after I'd purchased a flatbed scanner, thinking, "Hey, I can put all my Philo stuff on the Web!" Over the course of the next two years, I serialized my original text and published it online as *The Farnsworth Chronicles (http://farnovision.com)*. I'm pleased that the site has received tens of thousands of visitors, and that the legacy of Farnsworth's contribution to our daily lives has spread via this new medium, which is also predicated in part on his contribution.

It is no exaggeration to say that it has taken twenty-five years to write this book. When I first met the Farnsworths in 1975, and for many years thereafter, the family was quietly reluctant to talk much about Farnsworth's fusion work. It was not until I worked with Pem on the completion of *Distant Vision* in 1989 and '90 that some of the compelling details of those years began to surface. Still, there was not enough to effectively trace the arc of events and ideas that truly tell the story of this man's life and struggles.

An unexpected benefit rose out of the material I posted to the Web in 1998, when I created an online discussion board *(http://fusor.net)* and discovered that there is a small number of individual enthusiasts around the world who are experimenting with the work Farnsworth left unfinished in the 1960s. From the online discussions, I encountered Richard Hull, a "high energy" amateur experimenter from Richmond, Virginia, who has done his own extensive research into Farnsworth's last twenty years.

Richard's interest in fusion is equal to mine, and he had interviewed many of Farnsworth's co-workers from the fifties and sixties when our paths crossed. In the summer of 2001, I joined Richard for some follow-up interviews with Farnsworth's fusion team, which provided further insight into what was really going on at the Pontiac Street lab in Fort Wayne. In the past two years I have also found Pem and Kent Farnsworth much more willing to discuss Phil's fusion experiments. Thus, it has only recently become possible to reconstruct Farnsworth's last decade in such a way that much of the mystery surrounding his fusion work can be stripped away.

There are still many unanswered questions about the Fusor, and just what really happened while Farnsworth was with ITT. On the one hand, nobody can say for certain that the Fusor was ultimately capable of producing a self-sustaining fusion reaction or delivering a practical power plant. On the other hand, nobody can say for certain that it was not. What we do know is that the political and financial obstacles that Farnsworth faced were at least as daunting as the technical obstacles, if not more so.

Richard Hull has graciously allowed me to use here some of the material he has gathered, without which it would be impossible to complete the arc of the Farnsworth story that appears in these pages. Thanks to Richard's research, we now have a much better idea exactly what transpired during those "missing years."

Publishing my original material to the World Wide Web had one other unintended side-effect: this book. I first met Bruce Fries at an Internet music conference in 1999; when we encountered each other online again through an e-mail discussion group, I asked him to take a look at my website, to see if he thought there might be a book in there somewhere. When he said "yes," I was on my way to the culmination of twenty-five years of work. I fought Bruce a lot along the way, but in the end I can see that his suggestions were instrumental in strengthening and finding the true heart of this material. I am grateful to Bruce for his patience with

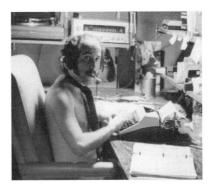

The author pounds out the first draft on a 70s "word-processor"

me, and also to Chris Roerden, the editor whose suggestions have gone a long way toward shaping this final manuscript.

I am immensely indebted to Kent Farnsworth for his tireless assistance in assembling the illustrations for this volume, and for his passionate fellowship over the past twenty-plus years. I am equally grateful to Kent's wife, Linda Farnsworth, for her personal strength and help through some of the challenging parts of the process. And I want to thank Georja Skinner, who has been a part of this effort from the very beginning, for her unwavering faith in this story.

Finally, I will consider it one of the great privileges of my life to have befriended Elma Farnsworth, and to have assisted her in whatever way I could in the preservation of her husband's legacy. I can only hope that Pem feels I have done justice to the full sweep of their story.

As the 75th anniversary of the first-ever electronic television transmission approaches in the fall of 2002, I expect there will be another wave in the mounting resurgence of interest in this man whose work so dramatically affected the course of our civilization. I hope that this volume, and all the years that have gone into its making, will add some texture and meaning to that celebration.

Paul Schatzkin

Pegram, Tennessee
May 2002

NOTES

Author's Notes

[1] The subject of this book is Philo T. Farnsworth II; his grandfather was Philo T. Farnsworth I, and first son, born in 1929, was Philo T. Farnsworth III.

[2] Einstein Announces the General Theory of Relativity—1915 WGBH/ PBS Science Odyssey: http://www.pbs.org/wgbh/aso/databank/ entries/dp15ei.html.

[3] David E. and Marshall Jon Fisher, *Tube: The Invention of Television;* 1996, Counterpoint Press, Washington DC, p. 29.

[4] Einstein's 1905 publications consisted of three groundbreaking papers. In addition to his Special Theory of Relativity, he produced a paper on Photoelectric Effect, which became the foundation for quantum mechanics. The third paper dealt with more arcane aspects of statistical mechanics. Source: The Albert Einstein Home Page: http://www.humboldt1.com/~gralsto/einstein/einstein.html.

[5] Elma G. Farnsworth, *Distant Vision: Romance & Discovery on an Invisible Frontier;* 1990, Pemberly Kent Publishers, Salt Lake City, p. 42.

[6] George Everson, *The Story of Television: The Life of Philo T. Farnsworth;* 1949, W.W. Norton, New York, reprinted 1974, Arno Press, New York, p. 38.

[7] Ibid., pp. 53–54.

[8] Farnsworth, *Distant Vision*, p. 54.

[9] Everson, *The Story of Television*, p. 54.

[10] Ibid., p. 55.

[11] Ibid., p. 63.

[12] Farnsworth, pp. 86–87.

[13] Ibid., p. 88.

[14] Ibid., p. 91.

[15] Accounts of the Ives/Bell Labs demonstration, with illustrations, and the resulting reportage in the *New York Times*, according to Fisher and Fisher, *Tube: The Invention of Television,* pp. 64–66; and Albert Abramson, *The History of Television, 1880 to 1941;* McFarland, Jefferson, NC; p. 99.

[16] In 1876 Alexander Graham Bell submitted his application for a telephone invention a mere three hours before another inventor, Elisha Grey, submitted a similar application. In the resulting litigation, lawyers

for Grey argued that their client had actually submitted his application before Bell's and that the examiner began his work from the top of the pile that had accumulated on his desk that day, reversing the applications' chronological order. Grey's attorneys were unable to win the case, or we would after all these years been using "The Grey System" instead of the "The Bell System" for our telephone communications.

[17] Farnsworth, p. 105.

[18] This account of the image magnification work—the precursor of the scanning electron microscope—comes from tapes recorded with Pem Farnsworth in 1975–76 (Schatzkin/Farnsworth Audio Archives); the "visitor" was from Westinghouse.

[19] The National Inventor's Hall of Fame credits three individuals with the invention of the Scanning Electron Microscope: James Hillier, Philo Taylor Farnsworth, and Heinrich Rohrer; http://invent.org.

[20] Farnsworth, p. 107.

[21] Everson, p. 114.

[22] *San Francisco Chronicle,* September 3, 1928.

[23] Farnsworth, p. 105.

[24] Account of Green Street Fire from Farnsworth, *Distant Vision,* p. 105, and Schatzkin/Farnsworth 1976 Audio Archives.

[25] Everson, p. 100.

[26] Ibid., p. 105.

[27] Ibid., p. 108.

[28] Ibid., p. 108.

[29] This depiction of Donald Lippincott is per Philo Farnsworth III, from Schatzkin/Farnsworth Audio Archives.

[30] Another depiction per Philo III, from Schatzkin/Farnsworth Audio Archives.

[31] Schatzkin/Farnsworth 1976 Audio Archives.

[32] Ibid.

[33] Any reader with more than a casual interest will want to learn more about Nikola Tesla, the enigmatic Serbian genius who created alternating current (AC), the method used to distribute almost all the electrical power in the world today. In advocating the use of alternating current, Tesla encountered fierce opposition from his former employer, Thomas Edison, who advocated the use of direct current in what became known as "The Battle of the Currents." Tesla eventually joined forces with George Westinghouse to build the first AC dynamo at Niag-

ara Falls, NY in 1896. Tesla also made seminal contributions to the then-new art of wireless. Suggested readings: Margaret Cheney, *Tesla: Man Out of Time*, Touchstone Books; Inez Hunt and Wantetta W. Draper, *Lightning in His Hand: The Life of Nikola Tesla*; Omni Publications; also recommended from PBS: *Tesla, Man of Lightning*; www.pbs.org/tesla.

[34] In his contract with Westinghouse, Tesla was granted generous royalties. But after the "Battle of the Currents" drained the company's resources, Westinghouse appealed to Tesla for relief, and Tesla is reputed to have torn up the contract out of respect for Westinghouse, who was the only industrialist bold enough to believe in his ideas. Westinghouse survived, but Tesla was left forever after in dire financial straits.

[35] *The History of Television, 1880 to 1941*; 1987, McFarland & Company, Jefferson, NC, p. 79.

[36] Ibid., p. 81.

[37] Marconi is generally credited with the invention of "wireless" telegraphy, the progenitor of radio. He founded the British and American Marconi Wireless Companies, the latter the forerunner of RCA.

[38] Frank Waldrop and Joseph Borkin, *Television: A Struggle for Power*; 1938, William Morrow and Co., New York, p. 167.

[39] RCA, by virtue of its cross-license agreements with AT&T, controlled the patents of radio pioneer Lee de Forest, among them a 1924 patent for the regeneration circuit, one of the cornerstones of radio reception and amplification. That patent was issued only after protracted litigation between de Forest and Edwin Armstrong (who most authorities consider the true inventor of the regeneration circuit), and was in force until 1941. Source: Tom Lewis, *Empire of the Air: The Men Who Made Radio*, 1991 Harper Collins, p. 204.

[40] Abramson, p. 123.

[41] Schatzkin/Farnsworth 1976 Audio Archives.

[42] Ibid.

[43] According to Tobe Rutherford's brother Romily Rutherford, who also worked for Farnsworth at the time, and reported in the PBS documentary "The American Experience: Big Dream, Small Screen" first broadcast in February 1997.

[44] Farnsworth, p. 132.

[45] Ibid., p. 133.

[46] Waldrop and Borkin, p. 181.

[47] Ibid., pp. 218–219.

[48] Donald G. Godfrey, *Philo T. Farnsworth: Father of Television;* 2001, University of Utah Press, Salt Lake City, p. 145.

[49] Ibid., p. 49.

[50] Farnsworth, p. 136.

[51] Schatzkin/Farnsworth 1976 Audio Archives.

[52] Everson, p. 135.

[53] Philco's experimental television station, W3XE, eventually became Philadelphia's NBC affiliate, and is still on the air as KYW-TV.

[54] Farnsworth, p. 143.

[55] Ibid., p. 143.

[56] Ibid., p. 144.

[57] Russ Varian returned to California and with his brother, Sigurd, and together they formed the very successful Varian Associates, one of the first companies to locate in what is now known as Silicon Valley. Russell Varian later patented the Klystron tube, one of the cornerstones of radar and microwave technology.

[58] This claim regarding Tihany's contribution to the Iconoscope was first introduced to the author by Tihany's daughter, Katarina Glass, in personal interviews in Los Angeles in the 1970s. The claim is substantiated by Abramson, p. 119; the text is accompanied by a diagram from Tihany's 1928 patent application, which shows a tube with a triangular configuration very similar to that of the Iconoscope.

[59] Schatzkin/Farnsworth 1976 Audio Archives.

[60] Farnsworth, p. 156.

[61] Ibid., p. 157.

[62] This second finding in the 1934 patent interference is critical because it states clearly that the device disclosed in the 1923 application could not have been the Iconoscope that emerged in the 1930s. See Appendix A, "Who Invented Television."

[63] Stephen F. Hofer, *"Philo Farnsworth: The Quiet Contributor to Television;* 1977, College of Bowling Green, unpublished doctoral dissertation, p. 81.

[64] Farnsworth, p. 147.

[65] Ibid., p. 150.

[66] Irving Berlin, *Always.*

[67] Farnsworth, p. 150.

[68] Ibid.

[69] Patent #2,246,625 "Television Scanning and Synchronization System" was first filed on May 5, 1930, but was not issued until June 24, 1941. For a complete listing of Farnsworth patents, see Godfrey, p. 189.

[70] 1976 interview with J.D. McGree in London, England; Schatzkin Audio Archives.

[71] *Newsweek,* August 10, 1935 p. 24.

[72] This is fairly typical procedure, i.e., applying for the patent before the device is actually built so that the work is protected as it is "reduced to practice." The patent in this case was #2,143,262 "Means of Electron Multipaction" filed 03/12/35 and issued 01/10/39.

[73] Everson, pp. 137–138.

[74] Ibid.

[75] Everson, pp. 138–139.

[76] Gerry Vassilatos, *Lost Science*; 1999, Adventures Unlimited Press, Kempton, IL, pp. 320–321.

[77] Paramount Eyes and Ears of the World, 1936.

[78] Collier's, *The National Weekly* October 3, 1936, p. 19.

[79] Farnsworth, p. 185.

[80] Ibid., p. 186.

[81] Everson, p. 159.

[82] *Business Week*, August 14, 1937.

[83] It's a little known fact that television's prestigious "Emmy" Award was named after the Image Orthicon Tube. When the Academy of Television Arts and Sciences (ATAS) was formed, former Farnsworth lab gang member Harry Lubcke was the Academy's president. Searching for a name as memorable as "Oscar" was for the Motion Picture Academy's annual awards, Lubcke suggested "Immy," borrowing from the name of the tube that was used in almost all the television broadcasts in the 1940s and '50s. When the trophy itself was unveiled, with the outstretched form of a woman holding the orbits of an atom aloft, the name of the statue morphed to "Emmy." Thus, every year, when the Academy bestows its annual awards for outstanding achievements, it does so with an unknowing nod toward the medium's seminal genius.

[84] Letter to The Stockholders of Farnsworth Television, Incorporated; August 9, 1937; from Schatzkin private archives.

[85] Godfrey, p. 256.

[86] Everson, p. 236.

[87] *Fortune*, May 1939, pp. 53–181.

[88] Ibid., p. 169.

[89] Ibid., p. 54.

[90] Ibid., p. 169.

[91] Farnsworth, p. 212.

[92] Everson, p. 246.

[93] Godfrey, p. 128.

[94] The most noteworthy, and disturbing, victim of Sarnoff's renewed resolve regarding patent royalties was Edwin H. Armstrong. In the 1910s and '20s, Armstrong invented circuits that were fundamental to the popular acceptance of radio, and enjoyed a profitable business relationship with RCA and a strong personal friendship with David Sarnoff. Responding to Sarnoff's offhand wish that somebody would invent a "black box" that would eliminate radio static, in 1933 Armstrong demonstrated a revolutionary new system of broadcasting—Frequency Modulation, or FM. Sarnoff at first supported Armstrong's FM work, and provided facilities at the Empire State Building to test FM's capabilities. But Sarnoff was wary of the impact that FM would have on the installed base of radio listeners, who would have had to junk their old AM receivers to switch to the new system. In 1935, Sarnoff curtailed Armstrong's experiments and evicted him from the Empire State Building facilities so that RCA could use the space to conduct experimental TV broadcasts. Sarnoff was not inclined to use FM as a new form of radio, but he did decide to use FM to broadcast the audio portion of television programs. He also decided to use Armstrong's patents without accepting the inventor's demands for continuing royalties, demands prompted in part by Farnsworth's success with his television patents. Armstrong litigated the case throughout the 1940s, exhausting his personal fortune. The case was ultimately resolved with Armstrong's widow—Sarnoff's former secretary—after Armstrong threw himself out of the 13th floor window of his New York apartment in 1954. Sarnoff's response on hearing the tragic news: "I did not kill Armstrong." References: Lawrence Lessing, *Man of High Fidelity: Edwin Howard Armstrong;* 1956, Lippincott; Eric Barnow, *A History of Broadcasting in the United States;* 1966, Oxford University Press, New York; Tom Lewis, *Empire of the Air: The Men Who Made Radio;* 1991, Harper Collins.

[95] Farnsworth, p. 216.

[96] Christopher H. Sterling and John M. Kittross, *Stay Tuned: A Concise History of American Broadcasting;* 1978, 1990, Belmont Wadsworth.

97 Farnsworth, p. 217.

98 Ibid., p. 117.

99 Ibid., p. 219.

100 Ibid., p. 223.

101 Ibid., p. 227.

102 Ibid., p. 226.

103 There is some precedent for such nomenclature. When experimenting with his earliest incandescent lightbulbs, Thomas Edison observed a current fluctuation that he did not fully understand, but the anomaly became known as the Edison Effect. Years later, John Fleming in England and Reginald Fessenden of Canada employed the Edison Effect in the first vacuum tubes, which were the foundation of all subsequent developments in electronics.

104 Schatzkin/Farnsworth 1976 Audio Archives.

105 Godfrey, p. 148.

106 Farnsworth, pp. 242–244.

107 Farnsworth, p. 249.

108 Kent Farnsworth was born September 4, 1948. He was the Farnsworth's fourth child, the third living child at the time of his birth. Second son Kenny died in 1933.

109 Godfrey, p. 211.

110 Farnsworth, pp. 266.

111 Godfrey, p. 208.

112 Farnsworth, p. 256–257.

113 Farnsworth, p. 257.

114 Ibid., p. 261.

115 Godfrey, p. 171.

116 Schatzkin/Farnsworth 1976 Audio Archives.

117 The first atomic bomb was tested in 1945 at Alamagordo, New Mexico; the first hydrogen (fusion) bomb was tested in 1952 on the South Pacific atoll of Eniwetok. The only two atomic bombs ever used as instruments of war were dropped on Hiroshima and Nagasaki, Japan, in August 1945 to end World War II.

118 The concept of electrostatic confinement was not original to Farnsworth, although he was the first to propose a device that would actually employ the idea. Writing about the earliest efforts to control fusion, former Atomic Energy Commission member Amasa Bishop wrote, "While not unequivocally ruled out, the use of electric fields for

confinement does not appear to be feasible.... Numerous proposals have been made for plasma confinement by electric fields, but none appears to have sufficient merit to warrant serious consideration." Instead, Bishop and others have advocated the use of giant magnets to confine a fusion plasma. Forty years and countless billions of dollars later we know *that* won't work, either. Source: Amasa S. Bishop, *Project Sherwood: The U.S. Program in Controlled Fusion;* 1958, Addison-Wesley Publishing Company for the United States Atomic Energy Commission, p. 15.

[119] The dates used in the *I've Got a Secret* appearance reflect some confusion regarding the actual dates of Farnsworth's conception. In this case, the onscreen caption showing "... in 1922, when I was 14 years old..." is incorrect. Farnsworth was born August 19, 1906. In the summer of 1922, he would have been 15 years old until August 18. Because we know for certain that Farnsworth was in Rigby High School during the 1921–22 school year, we fix the date of conception during the previous summer of 1921, when he would have been 14 years old. He likely disclosed to Tolman in March the following year, 1922, when he would have been 15 years old.

[120] Farnsworth, pp. 298–301.

[121] Ibid., p. 295.

[122] Accounts of the Pontiac Street activities by Richard Hull, postings to the Fusor.net website: http://www.Fusor.net/old-boards/intranets.com/history_news/msg-9631.html.

[123] This concern regarding the job security of the lab workers came to light during interviews with Gene Meeks and Steve Blaising in July 2001; Schatzkin Audio 2001 Archives.

[124] Farnsworth, pp. 296.

[125] F.R. Furth and R.L. Hirsch, *Unsolicited Proposal to Perform Basic Physical Research in the Inertial Containment of Ionized Gases;* 1967, ITT Industrial Laboratories Sec. 3, p. 2.

[126] Interview with R.L. Hirsch March 2002; Schatzkin Audio Archives.

[127] Farnsworth/Schatzkin 1976 Audio Archives.

[128] The sentiment that the Fusor experiments were "close" to their ultimate goal was echoed by Gene Meeks. Interviewed in July 2001, Gene said, "we were close...very close..." before the project was terminated in 1968. Schatzkin 2001 Audio Archives.

[129] Farnsworth/Schatzkin 1976 Audio Archives.

[130] Farnsworth, pp. 309–310.

[131] Blaising interview July 2001, Schatzkin Audio Archives; also Richard Hull, *End of Philo Farnsworth at ITT;* http://www.Fusor.net/old-boards/intranets.com/history_news/msg-9837.html.

[132] From a March 2002 interview with Robert L. Hirsch. Schatzkin 2002 Audio Archives.

[133] As this book is completed (May 2002) Pem Farnsworth is still very much alive, at age ninety-four, and living with her son Kent in Fort Wayne, Indiana.

[134] Individuals around the world who are experimenting with the Fusor share their experiences on the Internet at http://fusor.net.

[135] Fisher and Fisher, pp. 136–137.

[136] Abramson, p. 80.

[137] Abramson, p. 287.

[138] United States Patent Office; Patent Interference No. 64027—Farnsworth v. Zworykin—Television System. Final Hearing April 24, 1924, decision rendered July 20, 1935. The reference to "discrete globules" refers to a distinguishing feature of the Iconoscope, which Zworykin was testing in the mid-thirties. Since the decision says that the 1923 application did not disclose such globules, the decision further reinforces that the 1923 application was *not* the Iconoscope.

[139] Max Crosley, *Electromagnetic Spectrum Blues*, Radical Software *Videocity* Edition; 1973, Raindance Foundation, New York. I would have liked very much to publish the entire "Electro Magnetic Spectrum Blues," but Max Crosley is deceased and I was unable to obtain permission. However, the entire text can be read at the "farnovision.com" website: http://farnovision.com/chronicles/tfc-crosley.html.

[140] Philo T. Farnsworth III died in 1987 of cirrhosis of the liver.

Photo Credits

Pages 1, 7, 9, 11, 15, 20, 23, 29, 32, 35, 39, 44, 49, 52, 55, 67, 69, 72, 76, 79, 95, 107, 110, 118, 121, 122, 123, 125, 135, 137, 138, 139, 147, 153, 159, 161, 171, 181, 186, 187, 193, 199, 201, 203, 205, 211, 215, 223, 225, 227, 237, 239, 243, 245: Philo T. Farnsworth Archives.

Note: Photos from Philo T. Farnsworth Archives are now owned by the University of Utah, J. Willard Marriott Library, Special Collections Department.

Pages 83, 88, 93, 185: David Sarnoff Library.

Page 13: Nipkow disk. Unknown.

Page 14: Telephonoscope. Unknown.

Page 54: First electronic television image (recreated in 1977). Georja Skinner.

Page 56: Herbert Ives and the AT&T system. Lucent Technologies.

Page 95: Farnsworth's first patent. U.S. Patent Office.

Pages 128, 150: John Logie Baird. Unknown.

Page 225: Philo Farnsworth on *I've Got a Secret*. CBS.

Page 225: Moon landing/Farnsworth television set (composite photo). Paul Schatzkin.

Page 247: "Dessert cart" fusor. Steve Blaising collection.

Page 248: The "star" in Richard Hull's Fusor Mark III. Paul Schatzkin.

Page 249: Zworykin's 1938 Iconoscope patent. U.S. Patent Office.

Page 259: Paul Schatzkin with George Everson. Georja Skinner.

Page 263: Pem Farnsworth in 1977. Georja Skinner.

Page 264: Philo T. Farnsworth III in 1972. Kent M. Farnsworth.

Page 267: Paul Schatzkin. Self-portrait.

Page 285: Paul Schatzkin. Ken Gray.

INDEX

ABOUT THE AUTHOR

Paul Schatzkin was born in New York City during a hurricane. After graduating from Antioch College, Schatzkin gravitated toward Hollywood, where he served a number of years as a videotape editor on the ABC-TV comedy series "Barney Miller," for which he received an Emmy Award nomination. Schatzkin presently lives in Pegram, Tennessee

Author Paul Schatzkin in his 1949 Chevrolet pickup truck

with his wife Ann and their three cats. He drives a 50-year-old Chevy pickup truck and is engaged in an effort to resurrect the fusion research pioneered by Philo T. Farnsworth.

ABOUT TEAMCOM

TeamCom is a customer-focused publishing company that combines the best of traditional print publishing with new media, such as e-books and the Internet. Our mission is to deliver high-quality books, Web content and related products to help people improve and enrich their lives. For more information, please visit www.TeamComBooks.com.